How to Be a Researcher

How to Be a Researcher provides a strategic guide to the conduct of a successful research career within a university environment. Based on the author's extensive personal experience, it offers down-to-earth advice, philosophical guidance and discussions of the political context of academic research.

This is not a research methods book, and the topics it covers are rarely discussed elsewhere. The bulk of the book provides practical advice on the development of essential skills and strategic approaches, covering questions such as:

- how to decide which topics to work on
- how to read and review literature
- how to develop theory
- how to integrate research and teaching activity
- how to approach research design
- how to obtain and manage research funding
- how to collaborate and supervise effectively
- how to write up your research and
- how to secure the best sources of publication

The final part of the book considers the philosophy and psychology of research work and includes an exploration of the cognitive biases that may affect researchers.

How to Be a Researcher will be particularly useful for master's and doctoral students in the behavioral and social sciences, and also for early career academics developing research within a university career.

Jonathan St B T Evans is Emeritus Professor of Psychology at Plymouth University. He has over 40 years' experience of experimental research and has published more than 150 journal articles. He has also authored eight books, and he served as Editor of the journal *Thinking & Reasoning* for 17 years.

How to Be a Researcher

A strategic guide for academic success

Jonathan St B T Evans

Routledge
Taylor & Francis Group

LONDON AND NEW YORK

First published 2016
by Routledge
27 Church Road, Hove, East Sussex BN3 2FA

and by Routledge
711 Third Avenue, New York, NY 10017

Routledge is an imprint of the Taylor & Francis Group, an informa business

British Library Cataloguing-in-Publication Data
A catalogue record for this book is available from the British Library

Library of Congress Cataloging in Publication Data
Evans, Jonathan St. B. T., 1948–
    How to be a researcher : a strategic guide for academic success /
Jonathan St B T Evans. — 1 Edition.
        pages cm
    Includes bibliographical references.
    ISBN 978-1-138-91730-9 (hardback) — ISBN 978-1-138-91731-6
(pbk.) — ISBN 978-1-315-68892-3 (ebook)   1. Reasoning
(Psychology)   2. Research.   I. Title.
    BF442.E9294 2015
    001.4—dc23   2015009233

ISBN: 978-1-138-91730-9 (hbk)
ISBN: 978-1-138-91731-6 (pbk)
ISBN: 978-1-315-68892-3 (ebk)

Typeset in Times

by Apex CoVantage, LLC

Printed in Great Britain by Ashford Colour Press Ltd

Contents

List of Figures and Tables

Figures

Tables

Foreword

I have been engaged in academic research work for what seems like a very long time: in fact, since 1969, when I commenced my PhD studies. I am probably an unusual person, but I have found this a most satisfying and fulfilling way in which to spend my life. Over this lengthy period, I have learned a great deal about the process of research in ways that are philosophical, practical and personal. Very little of this was learned from books: certainly not books on research methods. The major stimuli for the current book derive from my efforts to convey this knowledge in supervising PhD students, mentoring junior colleagues and particularly teaching master's students studying Psychological Research Methods. I soon realised that what they most needed to know was not to be found in any available book. Instead, I had to teach them primarily from my own experiences as a researcher. So eventually I decided to write the missing book myself.

In 2005, I wrote *How to Do Research: A Psychologist's Guide*, published by Psychology Press, to cover these objectives. Of the books I have written, this remains my favourite. It seemed at the time to write itself, with little effort required on my part. I wrote it immediately after completion of a much heavier and scholarly work (Evans & Over, 2004) and on return (from sabbatical leave) to my normal university duties. Ordinarily, I would never have contemplated another book in such circumstances, but chapters of the current volume kept appearing unbidden in my mind, demanding to be written down. I am one of those writers whose work seems to incubate by unconscious processing and then flow on to the word processor. In one sense, the book required me to do no research at all in order to write it: in another sense, it had required 35 years to prepare it. Now 10 years later, I have decided to publish a much extended, revised and reorganized version of the book under the present revised title.

Looking back, I wonder why it took me quite so long to think of writing down what I had learned from my experience as a researcher in book form. I attribute this in part to the discipline of writing that had been instilled into me by years of professional research. What I – like all academic researchers – am obliged to write are *scholarly works*. These include conventional research methods texts. But what I needed to say could not possibly be expressed as a scholarly work. Such works do not allow their authors to write from personal experience, freely express opinions or to make claims that are not evidenced by published sources. But it became more and more obvious to me that you cannot teach someone how to write such a scholarly work or indeed how to conduct any other part of the strategy of research work by writing another such work. Why not then discard the writing habits of a lifetime and write a book that can do the job?

This book inevitably reflects my own experiences – and indeed the philosophy of science that I have developed over the years. I have tried to make it as broad as possible but there may be sections that are more helpful to those using an experimental approach and in particular to those engaged in more theoretical kinds of research. It is undoubtedly oriented to those wishing to build their research careers in the academic rather than the industrial environment, as the two are different in so many ways. I see potential readers primarily as ambitious young academics who may be at the master's or doctoral training phase or still finding their way in the early stages of a university teaching career. I hope also that it will be useful to advanced undergraduate students contemplating an academic career and wanting to get a feel of what is involved. We must all write what we know, so most of the examples of research discussed in the book reflect my training as an experimental psychologist. However, I also believe that the majority of the advice that I give is quite general and a good deal of it would be valuable to researchers in other disciplines, in particular across the social and biological sciences. I certainly had feedback from the 2005 edition from some researchers outside of psychology who had read the book and found it helpful.

When the earlier book was published, I did not envisage the need for future editions, as I believed the process of academic research and writing was unlikely to change very much. To some extent this was right, and as a result, I have left some substantial passages of text from the original book more or less unchanged. But I eventually realised that a new edition would be a good idea for two main reasons. First, the environment for academic research has changed in a number of ways. For example, there is far more publication of brief articles than there was 10 years ago, and online and open-access journals were in their infancy. There are also ever increasing political demands on academic researchers. For example, research must demonstrate real world relevance, to satisfy the funding bodies and their political masters, but also the highest theoretical value to secure the elite publications so craved by prestige-conscious universities. The second reason is that there were topics I could and perhaps should have covered in the original book but omitted to do so. I said nothing there about the integration of research and teaching activity, for example, and far too little about how to secure and manage funded research projects. Both of these topics have chapters of their own in the new book. Of course, I have also taken the opportunity to revise and update the original text throughout. In the new book, I also comment rather more explicitly on the internal and external political factors that affect researchers in their academic careers. I certainly did not understand academic politics very well early in my own career and wish now that I had done so.

I have also made one significant change to the organization of the book. It was intended primarily as an advice book: a guide for graduate students and early career academics on how to go about developing successful and fulfilling research careers in a university environment. However, I included two chapters in the middle that did something rather different: they explored the philosophy and psychology of research work. These were the chapters on hypothesis testing and reasoning and on statistical inference. When I came to revise the book, I realised that while these chapters include important information for early career academics, they do not (primarily) take the form of practical advice. As a result, I have separated them from the advice chapters (Part One of the revised text) and placed them at the end of the book in Part Two. Hence, all advice chapters (including two new ones that have been added) may be read through without any change of mindset being required by the reader. The focus throughout Part One is on the choices that researchers need to make and the skills they need to develop. For example: what topics to work on, how to read and review literature, how to develop theory, how to balance scholarship, teaching and research work, how best to benefit from collaboration, how to obtain and manage research grants, how best to write journal articles and other academic texts and how to go about getting your work published. I hope that Part Two will prove equally useful, but its somewhat different purpose is now made explicit. Here, I discuss the philosophy of science that underpins research work and also the *psychology* of science, in particular a number of cognitive biases that can affect researchers in designing their studies and interpreting their findings.

Although the objectives of the earlier book remain, the main title has changed to *How to Be a Researcher*. After publication, it became apparent that a number of potential readers saw the title *How to Do Research* and assumed (without reading it) that this was a book on research methods, similar to the many others on the market. In fact, his book contains little advice on research methodology, except incidentally. Instead, as the new title more accurately conveys, it is about everything that it takes to be a successful researcher in the highly competitive world of academia. The choices one has to make, the skills one needs to develop and the political context under which all activities from obtaining research funding to securing best publication of work need to operate. It is indeed a strategic guide for researchers.

It is hard to know whom to acknowledge for their contribution to my understanding of research work but I would like to mention my principal collaborators who are, in historical order, Steve Newstead, Simon Handley, David Over and Valerie Thompson. I learned something from all them as well as my PhD students and other collaborators. There were several readers of the draft 2005 book that precedes this one and whose comments continued to influence me in undertaking the rewrite: in particular, Robert Sternberg, Steven Sloman and Thomas Hefferman. I also thank Maureen Dollard for a critical reading of the draft of the new book with fresh eyes. Finally, I would like to make especial mention of my friend and collaborator, Shira Elqayam, who strongly encouraged me with this project and provided detailed readings of the manuscript for both the original book and the new edition published here.

Jonathan St B T Evans,
Plymouth,
February 2015.

Introduction

Numerous books on research methods written by and for psychologists are published every year. My bookshelf overflows with complimentary copies of such texts that publishers send me in the hope that I will recommend them to my students. I find the bulk of these books unimaginative and dull. A lot of authors seem to think that 'research methods' is little more than statistical analysis, based on psychologists' peculiar interpretation of statistical significance testing (see Chapter 9). The better ones tell you something about good design principles that ensure the control of extraneous or confounding variables. What is striking about all of them is what they do not tell you: how to do research. In a sense, they are much more about how *not* to do research.

As an analogy, imagine that someone teaches you the rules of chess, but nothing whatsoever about the strategy of the game. The function of the rules is to define some moves as legal and some as illegal: you may make the former, but not the latter. In early stages of chess games, there are typically around 30 choices of legal moves for either side. So armed with only the rules, you know which moves you must not make. However, nothing in these rules gives you any idea at all of which of the many legal moves available is a good choice. Without any understanding of the strategy of the game, you are sunk. This is the problem facing young researchers possessed of the latest texts on research methods and statistical techniques. They know the rules of the game, which tell them what they must not do (confound variables, for example), but this tells them almost nothing about how to do research. The purpose of this book is to cover the part that research methods texts leave out: the *strategy* of research.

My own pathway into research and academia was fortuitous to say the least. Having made a last minute decision to abandon the subjects of my A levels (mathematics and physics) and apply for a place in Psychology, I was an extremely lucky beneficiary of the selection system operating at University College London in the 1960's. They ignored A level grades (just as well, as mine were indifferent) and went entirely on their own selection procedure: a combination of psychometric testing and in-depth interviews. I was profoundly ignorant of academic psychology when I arrived at UCL and thought I was getting away from science. I coasted through the first year, doing just enough to get average marks. My whole subsequent life and career were shaped by a casual decision to attend a research talk given at the University of London Union by Peter Wason on the subject of negation. I had no idea what negation was and went along to find out.

I was fascinated by Wason's talk and had the temerity to knock on his door a few days later to ask him some questions about it. He strongly encouraged me to consider working with him as PhD student at the end of my degree, at a time when the idea of doing research or becoming a professional academic had never occurred to me. To cut a long story short, I did become his student, completed a PhD with him on the psychology of reasoning and have worked in and around the same topic ever since. I have no idea what has brought most of my colleagues into research but I have noticed that those fortunate enough to have done their PhD with a great psychologist, as I was privileged to do, seem to derive much advantage from it. There are many confounding variables, of course: great researchers tend to work in good departments, surrounded by strong research groups, supported by excellent facilities and with access to many international contacts. Nevertheless, there is a lesson here for prospective PhD students. Once you know what you want to work on, find out who does this work best and apply to work with them: do not worry about how big the name is.

Big names generally have big egos. If they publish prolifically, they are probably a little insecure, constantly trying to prove something to themselves and others. They are always flattered by someone taking an interest in their work. Some years ago, my daughter applied for a place in Mathematics at Cambridge. When she was called for an interview, she asked my advice on how to approach it. 'Ask them about their own research', I said. 'It never fails'. It did not: she got her place. Compared with famous people in most other walks of life, many big name academics are remarkably accessible. You want to ask them something: send them an email. More often than not, you will get a reply. As a young and totally unknown lecturer in the 1970's, I once wrote (no email in those days) to Noam Chomsky to ask him a question that had arisen in discussing his work in a tutorial. To my amazement, I got a lengthy and animated reply, spelling out his arguments in detail.

Career benefits of research work

Some people are very cynical about the motives of academic researchers. It is certainly true that research publications, in conjunction with your record of obtaining external research grants, are the main currency for purchase of career progression in academia, at least in the stronger universities. For example, if a young academic working in one university applies for a post in another, high profile university, then I can assure you that the prospective employers will not spend much time enquiring into their student ratings or administrative efficiency. I am not saying that good teaching and administration are not important aspects of an academic's life that should be given serious attention. Of course, they are. However, I am saying that the reality is that publication and increasingly external research funding records will be the dominant factors in securing promotion and esteem within the home university as well as fame and standing in the wider academic community.

The wealthier and more prestigious the university, the more this will be true. The stronger universities are not for those who want to devote most of their time and energy to teaching (but see Chapter 2).

Given this reality, successful academic researchers are often accused – usually by less successful ones – of being 'careerist': that is, publishing to advance their careers, rather than to advance scientific knowledge. The link between research, publication and career success is so strong that, inevitably, career considerations do have an influence, more so for some researchers than others. I think psychology has a particular problem with over-publication due to its popularity with students. So much student demand creates many large psychology departments around the world, all of which employ academic staff (or faculty) who need careers. The amount of work published reflects this, rather than the urgent need to solve important scientific problems. This can lead to duplication of effort, with similar fields that develop in isolation or ignorance from one another. It can also result in more or less deliberate ignoring and reinvention of work published 20 plus years earlier. The mass of publication also creates problems for scholarship (Chapter 1), as an individual can only absorb and integrate so much knowledge.

I am aware of cynical strategies that some researchers adopt for career advantage, but you will get no advice here on how to follow them! My focus is on how to advance knowledge and most effectively disseminate the findings. In general, doing good science will involve doing things that are good for your career at the same time. For example, you are not advancing knowledge in your subject if you do not publish it. Moreover, if you are able to publish in top journals, your work will be read by more people and have more impact in the subject. So although the advice offered in this book may help you advance your career, it will also help you advance science, as the two generally go together.

As an important qualification to the above comments, be warned that there are fads and fashions in academic research, meaning that some research fields are 'hot' or 'in' while others are not. These fashions can last for significantly long periods. I am well aware of this from personal experience. In the UK, for example, cognitive psychology (and more recently neuropsychology) has been largely dominated by research on memory, especially within the working memory paradigm of Alan Baddeley, for the best part of 40 years. In the first 15 years or so of my career, my interest in thinking and reasoning was regarded as idiosyncratic and clearly out of the main-stream. On an international scale, however, interest in the topic grew enormously in the 1980's and papers on the topic have appeared fairly regularly in the top journals from the 1990's onwards.

In spite of its expansion, the psychology of reasoning remains a minority interest in cognitive psychology and not the best vehicle for the ambitious. Somewhat fortuitously, however, I found that my interest in dual-process theory, which grew out of the reasoning work, eventually became very fashionable. I first published on this topic 40 years ago, but only quite recently have some of my publications on the dual processes attracted widespread interest and citations (e.g. Evans, 2003; 2008; Evans & Stanovich, 2013b). Needless to say, this happened too late in my career to have any practical utility! If you follow your interests, as I did, there is a big element of luck as to whether fame and fortune will follow. If you want a safer career route, then join in with the dominant paradigms in your field. But bear in mind that risky science can bring big rewards as well as failures.

Good careers that are principally focussed on degree level teaching are available in the USA, the UK and many other countries around the world. However, this book is aimed at those who wish to pursue academic research programmes, albeit typically combined with moderate teaching loads, in university environments. While I will talk about the best way to combine research and teaching to the benefit of both (Chapter 2), there can be little doubt of the importance of early and sustained success in research output in developing an academic career. In countries like the USA that operate a tenure-track system, strong research output is essential in order to secure long-term employment and equally important thereafter in securing promotion and salary advancement. In the UK, the introduction of selective research funding for university departments based upon the research output of their teaching staff has had dramatic effects on university politics and greatly strengthened the position of individuals capable of sustained international level research output.[1] In many European countries, the best academic researchers congregate in state-funded research centres and institutes. In North America, as in the UK, academic research is typically combined with a career that involves teaching university students, although there are also opportunities to pursue full-time research in some commercially sponsored laboratories.

The consequence of all this for early career academics is that they face tough choices in establishing their careers. For countries that use the tenure system, a good PhD and references may help you get a first job in a research-oriented university but further advance will follow the 'publish or perish' principle. Even the junior appointments in this system may be hard to obtain without good publications, however. Where tenure-track appointments are not an option (as in the UK, where employment law makes them effectively impossible), universities may be even more cautious about committing permanent appointments to unproven youngsters. Hence, one or more short-term postdoctoral research positions will almost certainly be required to provide a sufficient foundation of publications to secure a university appointment. So recall the advice about working with the best researchers. If you did not manage to get one as your PhD supervisor, try to secure one on your next postdoctoral contract.

The nature of research and researchers

What kind of person makes a good researcher? There are some general qualities that tend to predict success in a wide range of occupations: cognitive ability, motivation, good time management, etc. I will take these for granted and consider more the frame of mind that is needed for research. In my view, researchers fall broadly into two camps. I could call these theoretical and applied, but for the moment I will call them Scientist and Engineer. Scientists are people driven by a desire to understand nature: they want explanations and answers and may spend years or entire careers chipping away at some fragment of the great puzzle. Engineers, in my experience, although knowledgeable about science, have a different orientation. They want to solve practical problems in the world around them; they want to make things work.

Scientists are driven by the sheer joy of understanding and – in truth – are little concerned about the practical applications of their work. In psychology, those with this orientation tend to be drawn to topics that are highly theoretical in nature: for example, cognitive science, neuroscience or social cognition, despite the obvious practical implication that advances in these fields would bring. Those with an Engineering mindset are more likely to involve themselves with topics such as human factors, health and clinical psychology, despite the evident need for good theory in such fields. Of course, this is a simplification, as many researchers combine elements of both mindsets.

Having spent my career studying cognitive processes, I am clearly in the Scientist camp – although I hope that applied researchers will find this book useful as well. I personally find it impossible to observe almost anything in nature without wanting to understand how it works and where it came from. Every time I go on holiday to a new place, I return with a list of questions about the landscape for a friend of mine who is a Professor of Geology to answer. I am fascinated by weather systems and read books on meteorology for fun. I think physics is a really interesting subject. I lie awake worrying about whether the universe is open or closed. And I love philosophy because philosophers ask the *a priori* questions that must be addressed before the empirical study of science can even begin. Of course, not all psychologists are much interested in other sciences, but an almost childlike passion for understanding the world around you is a good indicator of the Scientist personality type. It depresses me how often the pre-school child's natural curiosity so often seems to be suppressed by exposure to formal education.

It can be difficult for a Scientist to answer questions from someone only interested in applications. For example, I am often asked about the practical implications of my theoretical work on thinking and reasoning, usually by people outside of the academic community. Of course, I can speculate about such applications – I have, after all, made many applications for grants to research councils. What people find hard to understand is why I have so little interest in personally developing them. They seem to assume that the *same* person would both do the science to develop understanding of an issue and then the engineering to apply it to a solution of a practical problem. If, indeed, they even understand that good science must precede application. On one occasion in the 1980's, we were visited by a member of one of Margaret Thatcher's right wing think tanks. He asked me to explain my research to him. He then interrupted me every few sentences, asking repeatedly, 'To what end? To what end?' Eventually, I ground to a halt, rendered uncharacteristically speechless. I had no idea what he wanted to hear; no theory of his mind at all.

Most psychologists, in my experience, read little or no philosophy and have only rudimentary knowledge of the philosophy of science. Yet, everyone who does research *has* a philosophy of science, whether they know it or not. One such philosophy can be unkindly described as 'brute empiricism'. This is an approach that sees science as the collection of facts through empirical observation. It is the approach criticised in Alan Newell's (1973) famous paper entitled, 'You can't play 20 questions with nature and win'. One of the early advocates of turning cognitive psychology into cognitive science was Phil Johnson-Laird. In 1977, he wrote '[Psychology] has an Empiricist obsession with experiments – with designing, executing, analysing, reporting and criticising them. In psychology, one experiment is worth a thousand theories'. The interesting point to note about such commentators is that they are not taking issue with their fellow psychologists about points of fact, and nor are they arguing the merits of one psychological theory over another. They are disputing the *philosophy* that underlies their colleagues' approach to research.

In general, I side with those who see the purpose of science to be the development of theory and I will devote a chapter (5) to this. However, empiricism is an important element in the philosophy of most scientists. That is to say, most believe that the *evidence* of science is based upon objective observations. What commentators like Newell and Johnson-Laird are getting at is their belief that the body of knowledge that constitutes science is theory rather than a list of facts. Good theories may be sorted from bad in part by their ability to account for the available empirical evidence. However, there may be other considerations. For example, theories should be coherent (lacking internal contradiction) and parsimonious (not making unnecessary assumptions). According to the philosopher Karl Popper (1959; 1962), scientific theories should also be falsifiable. Many psychologists advocate Popperianism, while practicing something that much more closely approximates to a Bayesian philosophy of science (Howson & Urbach, 2006). I will discuss these issues in Chapter 8.

Kuhn (1970) sees science as proceeding within *paradigms*. A Kuhnian paradigm is something much broader than psychologists' typical use of the term to refer to a particular empirical method. In paradigmatic science, researchers work with an agreed theoretical framework, set of objectives and methods in order to advance understanding. A good example in modern science might be the human genome project in the field of genetics. This provides stable, 'normal' science. Subjects yet to establish a clear paradigm are pre-scientific, according to Kuhn. Scientific revolutions occur when the current paradigm is challenged and overthrown. After a period of chaos and crisis, a new paradigm is established and normal science resumes. A dramatic example of scientific revolution was that initiated by Charles Darwin's theory of evolution by natural selection. This theory challenged not only strongly cherished scientific ideas but more widespread social and religious beliefs.

From a Kuhnian perspective, I think psychology has to be viewed as a set of disciplines. Compare, for example, cognitive psychology and social psychology. The cognitive paradigm was established by a Kuhnian revolution in which the previous paradigm of behaviourism that had dominated psychology for around 50 years up until about 1960 was largely rejected and replaced in an astonishingly short period of time. By the early 1970's, cognitive psychology was the established paradigm for the study of vision, memory, language and thinking within a theoretical framework based upon information processing. This revolution seems to have been inspired by the emergence first of cybernetic systems (Miller, Gallanter, & Pribram, 1960) and, subsequently, digital computers. The metaphor of brain as computer is very powerful, although strongly disputed by some philosophers (Searle, 1992). At some point in the 1980's, cognitive psychology merged seamlessly into cognitive science: the multidisciplinary study of intelligence, also involving such disciplines as philosophy, linguistics and artificial intelligence.

In psychology, new paradigms seem to emerge quite regularly. Where they appear to conflict with existing paradigms, this does not always cause a major crisis or revolution. Sometimes, they get assimilated. An example is the emergence of connectionism in cognitive science. When this first surfaced in the 1980's (McClelland & Rumelhart, 1985; 1986), it seemed to be in strong conflict with existing cognitive models of the day. Computational modelling of cognitive processes had relied upon symbolic, rule based systems that were explicit. Suddenly, we had neural networks requiring multiple interconnections between units and developing, through training, models comprised of numerous incomprehensible numerical weightings. Effectively, they were implicit models. Many psychologists were attracted to neural networks because they looked more like brains than rule based systems and solved problems (for example, pattern recognition) more effectively. However, the connectionist approach eventually extended the cognitive paradigm rather than overthrowing it. Some psychologists argue now for two distinct cognitive systems in the brain, one of which looks more like neural networks and the other more like rule processing (Evans, 2010; Stanovich, 2011).

Cognitive psychology currently has its own unresolved stresses, however. For example, neuroscience is highly fashionable at present and many cognitive researchers now feel pressure to demonstrate the relevance of their work to the brain (or vice versa). When I started out as a cognitive researcher in the 1970's, we could safely ignore the brain, although of course we knew it was responsible for the processes we were studying. It is impossible to ignore neuroscience these days, however, as studies using neural imaging or neuropsychological methods are published on every topic in cognitive psychology with increasing frequency. Having decided that if you cannot beat them, join them, I am myself collaborating on a major ESRC-funded project on the neuroscience of reasoning.

Like cognitive science, neuroscience is interdisciplinary but with the central focus on the brain function rather than cognitive processes. The extent to which you may think that these are two aspects of the same thing bring us back to philosophy: in this case, the philosophy of mind. Some neuroscientists see little place for a cognitive psychology that does not make direct reference to the brain. Some of us think in return that this is reductionism, a category mistake. Some happily merge the two paradigms and do cognitive neuropsychology. Another emergent and partially competitive paradigm is that of evolutionary psychology. Evolutionary psychologists view the structure of the mind as reflecting adaptations that slowly evolved from our ancient ancestors' interactions with the environment. They tend to emphasise the role of domain-specific cognitive modules within a massively modular brain. Arguments between evolutionary psychologists and orthodox cognitive psychologists take on a flavour more suggestive of paradigm conflict than of simple theoretical disagreement (Cosmides & Tooby, 1994; Fiddick, Cosmides, & Tooby, 2000; Fodor, 2001; Over, 2003; Stanovich & West, 2003).

In the course of this book, I will discuss not only the science of psychology but also – from time to time – the psychology of science (see especially Part Two). The latter concerns the cognitive basis of scientific thinking and the way in which beliefs, motivations and cognitive biases influence the conduct of our scientific work. My own specialist research field – thinking and reasoning – is helpful in this regard (I seem to have slipped into Engineering mode!). Psychologists in this area have studied much that is relevant to the psychology of science – the processes that underlie hypothesis testing and reasoning, for example (Chapter 8). There is also much work on the way in which we understand probability and statistics, some of which will be discussed in Chapter 9. In Part One, I will give practical advice on everything involved in research: scholarship, review, forming ideas and theories, designing empirical studies and obtaining the funding to run them, and – the most crucial topic of all – how to write up research and get it published. In Part Two, I deal with the philosophy underlying research and some psychological work that helps to understand the process of science and the cognitive biases that researchers may bring to it.

PART ONE

ADVICE

1 Scholarship and the origin of ideas

Scholarship – as I shall use the term – refers to the reading and interpretation of literature and the writing of literature reviews. Thus defined, you can be a scholar without being a researcher, but not *vice versa*. I am well aware that my old mentor, Peter Wason, would question my introduction of scholarship so early in a book on research, as he notoriously advised his PhD students to run their experiments first and read the literature afterwards (Evans & Johnson-Laird, 2003). Wason's concern was with creativity: he felt that if you spent too much time studying other people's ideas, it would inhibit your ability to produce your own. For the present, the point is that science is a collective international activity. You have to advance knowledge of the subject as a whole, not just your personal knowledge, so it is necessary to understand the relevant literature and to place your own work in this context. If you fail to do so, the journals will swiftly reject your papers.

Like good wine, scholars mature with age. I have discovered that one advantage (OK, the only advantage) of being an aging academic is that you accumulate knowledge across your career in a manner that puts the youngsters at a disadvantage. This is especially true if you work in the same field for many years, since you have read the literature as it happened, giving you a strong sense of the historical development of your subject. However, there is also clearly a danger of getting entrenched in one's ideas, as appears to be a general feature of aging, and young researchers are probably more likely to provide radical new thinking. Nevertheless, from the viewpoint of scholarship, young researchers clearly can only gain knowledge of their topic by reading the literature retrospectively. It is still important, however, to develop an understanding of how the field developed, rather than simply taking a snapshot of current knowledge, as represented in the most recently published papers.

This is where the modern method of literature searching is less than ideal in my view. Nowadays, everyone uses computer based search tools, allowing them to search on keywords or key authors. These are undoubtedly useful: *Current Contents* gives you a quick view of recently published work in journals of interest; the *Web of Science* (and similar tools) allows rapid search for literature by authors or keywords, as well as tracking citations. How did we do it before the search engines were available? I would start by scanning through the major journals that published the topic for, say, the past five years, copying any relevant papers. To discover the older literature, I would then examine citations in these papers, inferring the important sources from the frequency and nature of the references to them. The big advantage of this method is that it gives you the citation structure of the field and some feel for how things have developed. A related technique searches in the opposite direction, and search engines are very helpful here. Identify a key paper on the topic published, say, 20 years ago and then look at the papers published since then that cite this paper. Sometimes, you pick up papers you would otherwise miss because they were published in unusual places or lacked obviously linked keywords or titles.

While on the subject of computers, it is very important in the information age to understand the difference between knowledge and information. What is accessible by computer and, indeed, what is published in the journals is information. Knowledge is something that has to be constructed in the mind of the expert reader. This is what scholarship is about. Information is, these days, instantly accessible, but knowledge still takes years of dedicated study to acquire. Imagine that a freak accident wiped out an entire field of experts on a subject while all were attending a conference. How long would it take to reconstruct expertise in the field so that research could once again progress? It would probably take many years, despite the fact that their research was all published. To take another example, what do producers of science documentaries for television programmes do when they are researching their subjects? They talk to the experts rather than trying to read the journals. Quite rightly, as that is the only place that knowledge is to be found – inside the heads of the scholars.[1]

The purpose of academic research is to advance the collective knowledge of the discipline. Although the process also advances the knowledge of the individual, this is not sufficient. For example, if someone in ignorance of the literature 'discovers' something that was already known to the field, then they have not advanced knowledge in the sense that I am talking about here. However, things are really more complex than this. Research must be *contextualised* in its presentation. The context is the knowledge of the relevant topic prior to conducting your piece of original empirical research. In a journal article, this context is provided by the introduction, which includes a highly selective and focussed literature review. The discussion of the journal article relates the new findings to the context created. Hence, while scholarship is not sufficient to define research, it is necessary to conduct and report it. I will return to the writing of journal articles in Chapter 7. For now, I will consider the writing of a literature review as an object in itself. This might be a review written for the introductory chapters of a PhD thesis, for a review article, or for a book chapter or an entire book.

Writing a literature review

One of the reasons that experts frequently disagree is that no two people can ever have the same knowledge base. I am fortunate enough to have worked extensively with colleagues who shared a lot of my interests in thinking and reasoning. One of the reasons that research groups may be more productive than individuals (see

Chapter 6) is that there is a good deal of shared knowledge. Many of the papers that I have read and thought about will be known to my colleagues, so we do not have to waste a lot of time and effort establishing background knowledge when we want to discuss issues of interest to our research. This is, of course, facilitated by regular workshops and seminars. However, it is evident that no two people can have read exactly the same literature in the same order, nor have brought to it the same prior knowledge and attitude. If I am right in my assertion that knowledge, as opposed to information, exists only inside the heads of the human experts, then this implies that the knowledge acquired by science is not entirely objective and universal. It is to an extent subjective and individual.

Anyone involved in cutting edge research must be aware that the state of knowledge in their field is fuzzy and ill-defined. Leading authors promote conflicting theories and impose differing interpretations on the research findings reported in the literature. There *is* no definitive state of knowledge. It is important to be aware of this when writing a literature review. All such reviews, however scholarly, are constructions of the authors who write them. All reviews are by necessity selective in the citation and discussion of the massive amount of information that is available in the journals. No two authors would make the same selection or structure the material similarly, nor interpret it in precisely the same way. However, this does not mean that, as a scholar, you have licence to write anything you want. Some reviews will satisfy their academic critics (PhD examiners, journal reviewers, etc) and some most certainly will not.

The subjective aspect of a literature review became very clear to me when writing a review of my own field for the premier review journal *Psychological Bulletin* (Evans, 2002). The review encompassed a potentially huge amount of literature, since it covered the psychology of deductive reasoning over a 40 year period of work. The review was historical by its nature and purpose. The point was to argue that the central paradigm by which deduction was studied was based on a framework (logicism) that made sense when the paradigm was developed in the 1960's but that subsequent developments had made this methodology much more questionable as a method for studying human reasoning. The idea behind this paper was originally presented as a keynote address at a major conference and the reaction of colleagues to this was sufficiently positive to encourage me in the view that the field might be ready to hear this argument.

Note the references here to *purpose* and *argument*. Science is not the mere collection of facts, and nor is a review the mere summary and recording of evidence. You must always write to a purpose and indeed *read* to a purpose, as I shall explain below. However, the problem in writing a review for a journal, especially one as influential as *Psychological Bulletin*, is that you have to satisfy referees who are invariably other experts in the same field. Their view of the field will inevitably be different from yours but they know that your article, once published, will have a lot of influence on how other people – especially students and young researchers – will think about the topic. The first thing you have to do is impress these reviewers with your scholarship. You must appear to be (and actually be!) both widely read and possessed of a depth of understanding of the subject, even if it differs from their own. The second requirement they will have is that you are *fair* to other authors in the field – especially them! I will have a lot more to say about referees and publication in Chapter 7.

Literature reviews can be vastly influential. A good example is the book *Cognitive Psychology* (the first of many with that name), written by Ulrich Neisser (1967). Prior to publication of this book, the term 'cognitive psychology' was not in general use. Although the anti-behaviourist cognitive revolution had been in full swing for the previous 10 years or so, its replacement did not yet have a clear identity or name. Until Neisser's book was published, that is. From an objectivist viewpoint, Neisser did not invent the field – the research was already there in the literature, together with theoretical accounts reflecting the information processing approach. From the subjectivist point of view I am arguing, however, Neisser did in a sense contribute directly to the invention of the field. In this brilliant review, he constructed and identified the study of cognitive psychology. It influenced and inspired an entire generation of researchers, including me. The book was published while I was a final year undergraduate at UCL. It remains, to this day, the only psychology book I have read from cover to cover *twice*.

Structuring and organizing the review

Let us assume that you have done your literature search and have a pile of reprints sitting on your desk, or more likely these days, a folder full of pdf files on your computer. Your task is to read these papers and then write a review of them. How should you go about this? Most PhD students, in my experience, try to read the papers end to end, in no particular order. This approach is not only highly inefficient, but typically leaves the student exhausted and confused. Let us assume that most of these papers are journal articles reporting new empirical findings. (You certainly do not want to read someone else's review of the same field just before writing yours.) Each paper will tell its own story from the viewpoint of its own authors. Even though the papers fall into the same broad area, each is likely to have an introduction that reflects a somewhat different set of literature and a discussion that interprets findings within the authors' favoured theoretical perspective. Even papers that are presenting similar findings may frame and interpret them in quite different ways. Hence, if you read each paper within the structure that the authors provide, the total picture will be confusing and hard to integrate.

The key concept in literature review is *structure*. The review itself will need a structure, which can be done in several ways, as I discuss below. The sooner you can develop an outline of this structure, the better, as this will greatly assist you in the task of reading the literature. Hence, you need to do a preliminary reading that is relatively quick and superficial in order to decide how you wish to structure the review, and then do your detailed reading at a point where you have the main headings and subheadings. At this point, let me say a few words about depth of structure. Some PhD theses are presented using the numbering system of heading, which their authors take as a licence to embed to any level of depth. It is not unusual to encounter a section headed 2.1.3.4, or whatever. This is bad practice – do not do it! The point about your structuring of a review is that it must be reader-friendly. At all times, the reader must have your overall structure in mind and be able to follow where they are within it.

I strongly recommend an approach that allows only three levels: chapter, main heading and subheading. The first comes into play only when writing a thesis or a book. So a single paper, such as a review article, should permit only main and subheadings – just two levels. I have never encountered material that could not be structured in this way (although I should note that a distinguished commentator on my draft disagreed!). Keep in mind that the structure is something that you impose on the literature. You do it in order to make your assessment of it as easy as possible for the reader to follow and understand. Sections also have to be of an appropriate length that falls within the span of apprehension of the reader. As a general rule, I would recommend sections (or subsections) no longer than 2,000 words when reviewing technical material. If it needs to be longer, then divide it further.

The structure that you choose will, however, depend to some extent on the material that is being reviewed. I shall illustrate this with reference to a couple of my own publications. In the high level textbook on deductive reasoning that I wrote in collaboration with two colleagues (Evans, Newstead, & Byrne, 1993), we found that some chapters were best structured by the task that were used to study the phenomena: this worked well for Chapter 2 (conditionals) and Chapter 4 (disjunctives). But the same approach could not be used for the chapter (7) on syllogistic reasoning, as in this case, there is essentially only one task employed but a dozen different theories proposed to explain the findings. So that chapter was structured by theories, with major headings for related classes of theories and subheadings for specific accounts of the phenomena.

The second example is my review of dual process theories for the *Annual Review of Psychology* (Evans, 2008), my most successful paper in terms of citations. My remit was to cover such theories in three different fields: reasoning, decision making and social cognition. The second part of the paper does this, with a heading for each. However, my ambitions for the paper were more than to simply describe the major papers. I wanted to show a number of strands of common thinking that underlie dual process theories in these different fields and to identify common problems with the underlying thinking. So I started by listing similar attributes in the numerous theories of this type,[2] which I organized into four clusters: consciousness, evolution, functional characteristics and individual differences. Hence, the first part of the review had four subsections, allowing me to discuss the common issues with each of these clusters. Note that I did this *before* describing the theories in each field. Of course, my own understanding reflected in Part One emerged from reading the literature described in Part Two. But the reason for ordering it this way was to provide the reader with a frame within which more general issues were understood first and specific theories seen in that context.

It is important to identify as early as possible the structure that will suit your particular material. There is no uniquely correct way to do this, but a good way will reflect the various constraints that I have indicated: (a) it will reflect the nature of the material being reviewed, (b) it will enable you to make sense of the material and convey this understanding to the reader and (c) it will permit a two-level heading and subheading organization. Much of your detailed reading will be done *after* this structure has been decided. So how do you go about choosing the structure? This depends on whether you have some prior knowledge of the topic or whether you are coming to it cold. In either case, I would avoid reading other people's reviews of the topic, as you do not want to inherit their method of structuring the literature. If you are new to the field, then try to identify a subset of your mountain of reprints or pdf files that are the *key papers* in the field. Key papers are the ones that most influence research on the topic. Clues to which papers are key ones include the rate at which they are cited and the prestige value of the journals in which they are published. (Papers published in top ranked journals tend to be cited a lot, even if they are not particularly good.)

It is worth reading key papers in full, but with the others at this stage, you can just read the abstracts or skim the introductions in order to classify them. You should end up with heaps and sub-heaps of reprints (or folders and sub-folders of pdf files) corresponding to the headings and subheadings that you will use in your review. In this way, you can save much of your detailed reading to be done immediately before you write your 2,000 word section. This is very efficient, since if you read everything to start with, you will forget a lot and end up re-reading at the point of writing in any case. Even at the detailed reading and writing stage, it may not be necessary to read all the papers in full. It all depends upon the purpose to which you are reading and writing.

The writing of empirical journal articles is an art about which I will have much to say in Chapter 7. However, the basic schema has a lot in common with certain types of literature review. You have an introduction that lays out the basic context for the research; an experimental section that reports collections of new data and, finally, a discussion that makes sense of the new data in the context provided by the introduction. When I wrote my first major review – a textbook on deductive reasoning (Evans, 1982) – I already had about 10 years' experience of writing journal articles. What excited me about the review process was the much larger database of studies to interpret and discuss. I thought of each chapter as a big journal article, but instead of an experimental section with just two or three experiments of my own to consider, I had perhaps 20 to 40 studies on a particular topic to discuss. So what I did was read primarily the method and results sections of these journal articles, with minimal attention to the introduction and discussion sections. Where appropriate, related studies can be tabulated together, as we did quite frequently in the 1993 re-write of this textbook (Evans et al., 1993). You might even want to adopt the modern practice of meta-analysis (Rosenthal, 1993), where statistical techniques are used to provide inferential statistical analysis of a set of different studies.

Whichever technique you use, the point is that you are assembling data and then interpreting it. Reading the outer sections of the papers will hinder this process more than help it, as all the authors will have different ideas. It is *your* interpretation of the data that the review should reflect. However, that is not to say that introduction and discussion sections are a waste of journal space! You might, for example, be structuring a review section about theories, ideas and issues. In this case, the outer sections of the papers would be much more the focus of your attention than the inner ones. Again, it all depends on the purpose to which you are reading and writing.

The origin of ideas

Where do ideas for new research come from? This is an issue that philosophers of science have traditionally shied away from, declaring it to be primarily a psychological issue. There is much discussion to be found among the philosophers about how hypotheses and theories should be tested, some of which I will consider in Chapter 8. But where do they come from in the first place and to what extent does this depend upon scholarship? Recall that Peter Wason, who placed great emphasis on creativity in science (Wason, 1995), thought that reading and scholarship interfered with this process.

I think that we must acknowledge that creative researchers like Wason who lead thinking on their subjects are the exception rather than the rule. It is true that his work mainly reflected his thought about the issues rather than his scholarly appreciation of the work of others. A general problem is that the few sources of psychological studies of scientists and their thinking have tended to focus on great scientists and geniuses. Thus we have, for example, Wertheimer's (1961) account of the thought process of Albert Einstein as he grappled with the theory of relativity – fascinating stuff, but hardly helpful for the average PhD student. There are some systematic studies of great scientists to be found in the literature, such as the analysis of Alexander Graham Bell's diaries published by Gorman (1995). More useful, perhaps, are naturalistic studies of scientific research groups such as those conducted by Kevin Dunbar and his colleagues (Dunbar, 2002), but these generally provide more insight into reasoning processes than creativity.

'Normal' science is paradigmatic in the Kuhnian sense and hence does rely on scholarship to a large extent. To advance knowledge, we must know how far it has progressed in the first place. At the cutting edge, however, nothing is clear cut and decisive, certainly not in psychology. For every answer we find, we usually discover three new questions. In many ways, doing research is like reading a good novel. You get absorbed in the plot and eager to read the next chapter. Then you discover the next chapter has not yet been written, so you have to design your own research study to see where the story goes next. A student or early career researcher will usually get ideas from reading the literature, getting interested in an issue and starting to wonder about some aspect that is not yet understood. Or they may be inspired initially by an individual – perhaps a distinguished researcher in the department where they work. If you have the Scientist personality type that I discussed in the Introduction, you may develop a passionate curiosity that only research can satisfy. Established researchers may spend much time answering questions that their own earlier research has opened up, but there are individual differences here to which I will return.

Creativity in research is hard to pin down but one can make some general observations about it. We know as a general principle that creative thinking requires individuals to feel free to generate a lot of ideas uncritically. Of course, they must be critically examined after generation (and the majority rejected) but an unduly critical attitude by research supervisors – which may become internalised by the student later – is not helpful. Wason (1995), writing on this topic, made several interesting observations. Research supervisors (see Chapter 6), he tells us, should show most enthusiasm at the *start* rather than end of a project: 'I frequently used to bombard my students with letters if I thought there was anything I could add to their ideas. An idea may be 'half-baked' . . . but . . . given careful attention, can become 'baked' in the course of time rather than burned' (p. 290). In common with the anecdotes recorded from other creative scientists, Wason comments on the fact that ideas may become suddenly conscious when attention is on other matters. The key here is to recognise and pursue such moments of inspiration.

New ideas often come from synthesis of different, apparently unconnected pieces of information. I had the opportunity to write a personal history of the development of the dual process theory of reasoning (Evans, 2004), an enterprise in which I had been involved for more than a quarter of a century by that time. A landmark publication along the way was that of Evans and Over (1996) in which we started to develop the idea of dual systems underlying dual processes. A major influence on this was fortuitous to say the least. I was asked to review two books on implicit learning, whose authors made no reference whatsoever to the psychology of reasoning. However, they discussed a theoretical distinction between implicit and explicit cognitive systems that mapped uncannily on to ideas that I had been developing about reasoning and filled in many theoretical gaps for me. In the same article, I acknowledge an unconscious influence on the development of the modern theory. I published for several years about the distinction between two kinds of rationality before realising that I had been strongly influenced by a paper in *Behavioral and Brain Sciences* by John Anderson. In fact, the idea was essentially expressed in a commentary that I actually published on Anderson's paper at the time and then completely forgot that I had written!

Academic research is a very public business. We publish in journals available to all, meet up at conferences to exchange our most recent ideas, debate issues with colleagues around the world in email messages and spend hours discussing issues with the immediate members of our research groups. One consequence of this is that the origin of specific ideas may be hard to pin down. Occasionally, one gets into disputes where typically a research student or junior colleague accuses a senior colleague of stealing an idea (see Chapter 6). They may not be aware that their research supervisor could have deliberately planted some of their 'own' ideas in the first place, to help build their confidence. My experience is that research groups work best when people do not try to defend individual ownership of ideas but accept that they evolve through a group over time. As long as authorship of papers includes all relevant contributors, this rarely causes a problem.

I find that experienced researchers generally fall into two basic kinds: opportunistic and programmatic. Some people research a topic for a few years at most, publish a few papers and then move on to something else. Others carry out research programmes that span many years or even their whole career. I confess to being of the latter, programmatic type. In the past decade or so, I have published mostly on conditionals (e.g. Evans, Handley, & Over, 2003; Evans & Over, 2004) and dual processes (e.g. Evans, 2008; 2010; Evans & Stanovich, 2013b), topics that engaged my interest from the early 1970's onwards. I have run a secondary line of work on probability judgement that dates from the mid-1970's, but even this is closely linked to the reasoning work – at least in my own mind. I have returned to the same issues again and again over the years, each time making some progression (or so I like to think). I know a number of colleagues, however, who would blanch at the prospect of working on the same issues for more than five years or so. They bore easily (so they say) and get interested in some new paper they came across, perhaps in order to prepare an undergraduate lecture.

In truth, it is very hard to know precisely how much you are influenced by what you read in the journals or hear at conferences. This is an area where our self-knowledge is notoriously poor. Social psychologists studying attitude change have shown that we may not even know what we believed a couple of weeks ago, so having any accurate meta-cognition of thought processes that develop over months or years is a pretty hopeless prospect. It is as well to play safe when crediting others. Cite authors who have had similar ideas, whether or not you think they influenced you. Make sure that all colleagues and students involved with a research project are included as authors on your paper or explicitly acknowledged in a footnote and so on. No matter what you do, the occasional referee will accuse you of intellectual robbery. I was once accused by a journal reviewer with an agenda (all too common) of lifting ideas from an unpublished manuscript that I had never seen! It was a chapter sent in (too late) for a book being edited by a collaborator of mine. The author/referee just assumed that it had been passed on to me (which it had not) and was infuriated by coincidental thinking in a later article of mine. Fortunately, I was able to show that the idea in question had been published by me in a previous source (unknown to the complainant) that *pre*-dated their own chapter by several years.

Parallel discovery is an established phenomenon in science. I have become aware of this on a number of occasions by refereeing manuscripts sent to different journals that were obviously reporting independently designed research discovering essentially the same effects. I have also been involved personally on several occasions. Sometimes, a published article in a high status journal has a rather obvious flaw in its method or argument that somehow eluded the original referees. Leading researchers are irritated by this and design some new experiments to demonstrate the problem, only to discover that someone else has published the same critique first. More perplexing are cases where there is no obvious common factor, such as a recently published paper. It sometimes just happens that research groups working on the same topic area decide to run the same novel experimentation at the same time. We have quite often had discussions in my own group when we think of a new line of experimentation in which we say, 'What an obvious experiment. Why hasn't anyone done this yet?' Sometimes, it turns out that they are – as we speak!

This phenomenon is comforting really, even if it can be an annoying experience for the participants. If science is the collective advancement of knowledge, then it should happen that a particular phenomenon is ripe for discovery at a particular time. Otherwise, research advancements would depend upon the particular individual researchers who are around at a given time. If there is an objective (as well as subjective) basis to scientific knowledge, then the truth should out at some point. Sir Isaac Newton discovered the system of mathematical analysis now known as 'calculus' but kept it to himself for years in order to have an advantage over his rival scientists. He was then enraged when Leibniz later published the system, convinced (erroneously) that his idea had somehow been stolen. Other great ideas such as relativity theory or natural selection would surely have been with us by now, even if Einstein and Darwin had died at birth.

When I undertook my own PhD work in the early 1970's, it was not uncommon for students to receive little support from their supervisors. I remember one unfortunate colleague being told on his first day to go to his room and not come back until he had thought of an idea. I knew another who had his registration upgraded from MPhil to PhD by a supervisor whom he had never met. A US based colleague contributed a similar American horror story from the 1970's: the case of a supervisor who promised to pass a dissertation provided he did not have to read it! Fortunately, such days are long gone (see Chapter 6 for discussion of PhD supervision). The UK research councils, which fund such students, cleaned up the act in the 1980's by obliging universities to ensure completion of PhD theses within a given period of time as a condition of continued funding. Completion rates soared as a result of proper intellectual support being provided. In the USA, there is also much greater accountability for supervision nowadays, due to pressure from both promotion committees and funding agencies. No research student these days should flounder for want of an idea being given to them (consciously or unconsciously) to get them going.

KEY POINTS

- Academic research is the advancement of knowledge for all, not just the individual
- All new research must be presented within the context of what is already known in order to show how it advances knowledge
- Scholarship is an essential skill, comprised of the ability to read, assimilate and make theoretical sense of a wide range of academic literature
- Knowledge is to an extent subjective since no two individuals can share the same reading and understanding. There is no definitive state of knowledge but books and review articles can help to create a broad consensual understanding

- Literature reviews are not simply descriptions of known facts. They must be written to a purpose and structured and organized accordingly
- Literature should also be *read* to a purpose with the structure and focus of the eventual review article formulated as early as possible
- The origin of ideas is something of a mystery but reading of literature and conduct of empirical studies will usually engender them

2   Research and teaching

Universities are in the knowledge business: creating it (research), assimilating and organizing it (scholarship) and disseminating it (publication, teaching). The tradition of students grouping around expert scholars to acquire their wisdom has been with us for hundreds of years and shows little sign of changing. It might, of course, given the technologies available these days. Some future politicians or bureaucrats might decide that it is an unnecessary cost for students to attend many different universities for personal teaching when the curriculum could be delivered nationally or internationally by recorded lectures and distance learning materials. It will be a sad day if it comes, however. I believe that the personal contact between students and those expert in and passionate about their subjects adds something invaluable to the process of education.

For the time being, at least, most academic researchers will spend a significant part of their working life teaching students, with all the associated work such as lecture preparation, marking or grading assessments and attending relevant committees. Full-time research is a privilege that few of us will sustain for any long-term period. Teaching is a major part of the work of a university and an essential source of income. Many, but by no means all, researchers enjoy this side of the job. Whatever their attitude, however, teaching presents a problem for research work because it is so demanding of time and effort. It is not an option to neglect teaching or do it poorly. Even if your professional pride would allow it, students will complain, colleagues will notice and your career will suffer. A successful academic researcher will need to find a way of teaching well and being productive in research at the same time.

In the context of higher education, do good researchers make good teachers? Throughout my career, I have heard this debated many times. Within higher education, there seem to be two conflicting views, although the balance between the two camps will vary greatly according to the nature of the institution. As elsewhere in this book, I shall caricature two extreme positions. The Researcher theory proposes that researchers are smart, expert in their subjects and all-round good at their jobs. Hence, they make the best teachers. The Teacher theory says that researchers are self-centred, career-minded and neglect students in order to spend more time generating career enhancing publications. Hence, they are worse teachers than those who dedicate their working time mostly to their students. As an individual researcher and also at one time as head of a highly research active school of psychology, I have often had to make defences against critics who adopted the Teacher theory.

There is truth in both theories, of course. As a famous psychometrician once said to me, of human attributes, 'all good things are correlated'. The most productive researchers in a department are often also more able, hard-working and interested in their subjects than their less productive colleagues. Many of them apply these attributes to the benefit of their students and are very good teachers. However, I have also met researchers who do seem to want to pursue research at the expense of students, if they can get away with it. I have heard teaching described as a 'punishment' and known some senior researchers declare themselves to be too important to teach. I have also noticed that such individuals rarely generate enough research income to come remotely close to paying for their salaries! Such faculty create an equal and opposite problem for their universities than the staff who focus all their attention on teaching and neglect research work. As a head of school, I had to deal with both kinds of problem on a regular basis.

In favour of the Teacher theory, there are certainly individuals who spend a great deal of time and effort on teaching and do a very good job of it. If they have let research go, however, it is not usually because they cannot find time for it. Sometimes, people will do very good PhD and perhaps some postdoctoral work but once they obtain a teaching post, their research falls away. Depending on whether or not they work in a tenure-track system, the university may be obliged to employ such individuals for a very long time. I have given quite a lot of thought to why this happens and I do not believe it is generally due to excessive demands of teaching or lack of motivation of the individuals concerned. After all, in most universities, excellence in research work will bring great advantage in promotion and salary prospects. Rather, I think the problem is that some people can do good research work in a supervised environment, but are not capable of driving their own research programmes. They just lack an essential element of creativity. This is something of which appointment committees need to be aware, but that is rather beyond the scope of the present book.

For my target audience, the ambitious early career academic researcher, the situation is essentially this: you will need to spend a lot of time and effort on teaching students, while at the same time developing a strong research programme. For the rest of this chapter, I consider how these two tasks can be most effectively combined.

Time management

I have not noticed any strong correlation between the number of hours that my colleagues work and their productivity. I have, however, noticed that some people are much more efficient in their use of time than others. A lot of the advice I gave in Chapter 1 concerned efficient use of time in studying literature and writing reviews. There are certainly much quicker and slower ways of completing the same task. Part of the problem of combining good teaching with productive research work comes down to efficiency and time management – in both domains.

The reading that young researchers engage in prior to their first teaching appointment will often be quite narrowly focussed on their research topics. As with many topics discussed in this book, this will vary with the country and system in which people are working. Where there is a graduate school structure with mandatory higher level taught courses to attend alongside PhD work, researchers may emerge with a broader range of knowledge. In most cases, however, those confronted with preparing their first lecture course from scratch will be shocked to discover how much they do not know. Of course, there is far more to good teaching than having accurate knowledge to pass on. You also have to select and organize the material and present it in a comprehensible manner to your students.

Much of the advice that I gave about reading and reviewing literature in Chapter 1 can be applied to lecture preparation as well. You need to identify the structure of the lecture and the key research papers as soon as possible, so that you are reading and writing the lecture to an organized purpose. If you read the literature for your lecture in an unstructured and undirected manner, you can suffer exactly the same costs that I described there. You will take a very long time and become confused in the process. You will also become anxious because if you cannot understand the material, what chance do your students have? In this case, the purpose to which you are reading and writing is the teaching of the material. So decide as early as possible what your students need to know, how much they can take in within the time available and the easiest structure for them to follow.

Most young lecturers prepare far too much material. That is ok as long as you learn from experience and do not keep repeating the same mistake. Over-preparation wastes time you can ill afford and does not benefit your students – especially if you then try to cover it all in the lecture! I think many if not most academics are prone to anxiety and obessionality. These are traits that can be harnessed into good productive work but that can also be very destructive if allowed free rein. It is vitally important to know when you have done enough and stop. This applies equally to writing an undergraduate lecture and preparing a paper for submission to a journal. You must be properly prepared, but if you allow yourself to be too obsessional, then you will always need one more draft, to check out one more reference, and so on. You must balance the understandable anxiety you feel at the prospect of a poorly received lecture or a rejected journal article against the vital need to use your time efficiently. Finding the right stopping point is all important in this. If you are very inexperienced, then seek the advice of experienced colleagues who appear to manage their own time well. But it is important to develop your own sense of when to stop at the right time. Endless iteration is the enemy of efficiency.

When preparing lectures and other course materials, it is also good to think long term. You will most likely teach this subject again many times in your career. You obviously cannot protect and guarantee your lectures against future developments in the subject. The whole point of being a university teacher is to keep your students up to date with the latest research work. However, you can design your materials with this in mind so as to make regular updating easier to carry out. And when you do repeat earlier teaching, the same warnings about anxiety and obsessional behaviour apply. You do not need a major overhaul every year, particularly when teaching at basic undergraduate level. You do, of course, need one from time to time.

Time management is not just about efficiency, it is also about *scheduling*. When you have a number of tasks to do of differing urgency, you cannot give priority to the urgent tasks all the time. Imagine if a hospital did this on receiving patients for emergency treatments. Yes, they will treat really urgent cases immediately, but the least urgent still have to be treated at some point or people would be waiting forever. And here lies a problem in balancing teaching and research work. Teaching is almost always more urgent. You have a paper you want to write as soon as possible, but that lecture has to be delivered tomorrow and those papers have to be graded by Friday and so on. So research work can be put on the back burner for far too long, if you are not careful.

The approach to this problem that I recommend is to blank out certain days or half days (that are clear of teaching) for research and stick strongly to that plan. That time is simply not available for other tasks. Your planning of urgent teaching and administrative tasks has to work around this. For example, if Thursday is a research day and you have a lecture to give on Friday morning, then the deadline for preparing that lecture is Wednesday evening. I worked in this way most of my career so I know it is quite possible. It does require a lot of self-discipline, however. And, of course, these precious research days are not to be wasted. So it is good also to plan ahead what you want to achieve and fit the task to the time available. It is not a good idea to set aside research and writing altogether during a busy period if you can possibly avoid it. You will lose the thread of your ideas and have to start again – a most inefficient process.

Common elements in teaching and research

It would be ideal simply to teach what we research. In practice, this will not happen because (a) the teaching requires a much broader and shallower coverage of material than research work, and (b) the department will need to deliver particular courses and may not always have specialists available. In my own case, my research has been almost entirely within the psychology of thinking, especially on reasoning and decision making. However, at various times in my career, I have been called upon to teach a variety of other topics including perception, language, mathematical psychology, history and philosophy of psychology and even social psychology. I have also taught research methods quite a lot, but that is quite natural, as good researchers need to have detailed understanding of methodology. In fact, while some of these courses were a pain to prepare from scratch, they all advanced my knowledge and understanding of psychology in different ways.

At the very least, I was enhanced as a scholar by doing this and, in some cases, unlikely connections were made that gave me ideas for new research projects.

Most people get the chance at least to teach the topic areas within which their research interests lie. Of course, these are usual broad topics like memory, visual perception or social psychology, whereas your research will focus on something much more specific: phonological access to memories, face recognition or attitude polarization. But such breadth of knowledge is good and should be welcomed. It gives you a perspective within which you can understand your own research work better. And the reading you do for your teaching will surprisingly often provide ideas and theoretical insights that will directly benefit your research.

The extent to which teaching will directly stimulate research will depend to some extent on your research style. As discussed earlier, there are different research styles. Some researchers are programmatic, pursuing related issues over long periods, sometimes entire careers. Others are more opportunistic, following up ideas of interest as they arise, without any necessary overall connection. Opportunistic researchers are more likely to follow up ideas that they derive from reading something for the purposes of teaching. As a programmatic researcher, I have had less direct benefit of this kind but still quite a lot overall.

As an example, in the early days of my teaching career, I taught mathematical psychology, considered an important subject at that time (mid-1970's) but soon to go somewhat out of fashion (but making a strong comeback in the past decade!). I had some background in mathematics and had taken a special option on mathematical psychology in my undergraduate training. I taught topics such as mathematical learning theory, including Markov models, signal detection theory and stochastic choice theory. (It was hard work, given the resistance of the average psychology student to technical topics.) I learned some interesting things from this, such as how two different models based on completely different psychological principles could account very well for the same data set. This teaching fed into my own research to the extent that I published a stochastic model of reasoning (Evans, 1977). The paper was largely ignored at the time, as flow chart models were by then much more fashionable. But mathematical modelling is back now in the psychology of reasoning and that early paper was eventually followed up using much more sophisticated modelling techniques (e.g. Klauer, Stahl, & Erdfelder, 2007). The 1977 paper also provided the conceptual basis for a much more recent paper of mine that has been well cited (Evans, 2007b).

If you remain in the same department for some years, opportunities will arise to be proactive by proposing courses that you would like to teach. By this means, you can choose to do some of your teaching directly on your own research field, and sometimes include your own research papers. I was a bit hesitant to do this at first, but have generally found a very positive response from students over the years. They like the feeling of being near the cutting edge and being taught by someone who actively participates on the topic. Of course, if you are a major player in the topic you are teaching, it would be impossible to give it balanced coverage leaving your own work out. But 'balanced' is the key word here: you cannot teach that you are right and everyone else is wrong!

Actually, the problem that arises here is similar to that of writing a major review paper on a topic that includes your own research work. You need somehow to review your own work as though it was written by someone else and on the same basis as other authors. Not easy to do, of course, but journal referees (and students) will not allow you to take an obviously partial approach. When teaching my own work, however, I do take the opportunity to explain to my students how the work came about and the thinking that underlies it. I think this provides valuable insights for the students on the process of research. The conventions of writing journal articles (see Chapter 7) are such that the process is generally heavily disguised, sometimes being contextualised in introductory sections by studies that were published after the work was completed. The appearance that one piece of research is elegantly deduced from those that preceded it is, of course, an illusion! Here is a chance to tell your students how research really happens. I also try to humanize the process when teaching the research of others, especially if I know them personally and can recall some relevant anecdote about them.

If you are the person who will mark or grade the student's work, then there is an ethical issue that arises when you include your own work in your teaching. I would always tell my students explicitly to treat me as any other author and forget that I am marking their essays. In particular, they should know that they are expected to assess the work of all authors critically. They were definitely not going to get marks for uncritical praise of my own papers! This may sound obvious, but you do need to spell it out, as students may quite understandably worry about how they should do this. In practice, I never had a problem with this. I included coverage of my own work many times in my long teaching career and this was never once the cause of a complaint from a student.

Undergraduate research projects

The nature of undergraduate research projects varies greatly with teaching methods in different countries. I can only report on my own experiences within the UK. Here, the system is that students undertake a major (for them) project in their final undergraduate year, counting the same as several taught modules. In practice, this generally enables them to conduct, analyse and report a single psychological experiment, large enough to test for significant effects. Many studies are beyond the scope of an undergraduate project: for example, psychometric studies requiring large samples. Some topics that students might like to cover – in clinical or developmental psychology, for example – may be ruled out on practical or ethical grounds. All projects must be supervised by a member of the teaching staff, both to provide expert guidance and to ensure that the work reported was actually carried out.

Even within the UK, there is great variation among psychology departments in how student projects are organized and assessed. One model, favoured by Teacher theory, is that it is an opportunity for students to be creative and work on a topic that interests them. Another, favoured by Researcher theory, is that students are apprentices learning their trade under expert direction of the staff.

The problem with the first approach, in my experience, is that many students lack the creativity and imagination to generate their own projects. An approach that is too laissez-faire will result in many *ad hoc* and poorly designed surveys being conducted in the student bars. The danger of the second approach is that faculty may see students as unpaid research assistants simply there to carry out their own research work. If over-directed, the opportunity to credit students with originality and creativity will be greatly diminished. Over the years in my own department, this balance was the cause of many arguments and internal policy papers.

My own approach was always to encourage students to generate their own ideas and research designs where possible. In the majority of cases, however, it was necessary to direct them towards something specific. This would be enough to set some of them well in motion, while others would need continuous help at each stage. It was always considered acceptable to publish papers based on student projects from time to time and I did so from an early stage in my career. But not very many, as we shall see. This is another area where policy issues arise. In my own department, we eventually developed explicit, written policies on the publication of student projects. First, no paper for publication could be written before the student had completed the process of submitting their project report, having it assessed and the degree awarded. This avoided possible interference with the assessment process. An early draft framed by the supervisor would be of great advantage to a student yet to submit their project report, for example. Even if withheld from the student, the writing of such a draft could colour the supervisor's view of the project and make it harder to separate their own input from that of the student.

The other rules we followed were these: projects could only be published with the knowledge and agreement of the student; the student must be offered the opportunity to write the paper themselves if they were willing and able to do so (in practice, this almost never happened); finally, the paper could be written by the supervisor as first author but the student's name must also appear on the paper even if they contributed little to the design and/or had no ambitions for a research career. I cannot guarantee that these rules were followed by everyone but I can say that I have followed them myself throughout my career.

It is evident, then, that the supervision of undergraduate projects may provide an opportunity for the supervisor to advance their own research work. At the very least, an idea could be piloted supporting later research work or to help formulate a grant application. Conceivably, if results are clear cut enough, then the project might be published as a short journal article in its own right, or together with other experiments as part of a larger paper. In my own career, I estimate I have supervised around 250 undergraduate research projects. I am notorious for rarely missing a publication opportunity, so the reader might be interested to know how many of these were published. The answer is six. It is easy for me to be precise, as I only have to look through my list of publications and pick out those of the form Evans and X, where X is a name that never appears again. In any case, I remember the projects even if I cannot always picture the student.

Why so few over a period of 40 odd years? First of all, because students learn by making mistakes and do not always manage their time well, many projects do not deliver experiments of publishable quality. There may be an insufficient number of participants, errors in conduct and analysis, and so on. Where students choose to pursue their own topics (always the first option under my supervision), they rarely produce a publishable piece of work and I would not be writing it up in any case. The other reason that I was very cautious about publishing student projects was that I needed to be completely sure of the integrity of the student. Not only did I have to guard against unintentional errors but also deliberate falsification of data. It is a sad fact that various forms of academic dishonesty occur quite frequently among students. I always put my academic reputation first and was never willing to risk publishing an unreliable finding. In short, not only did the design and results have to meet publishable standards, I also needed to be convinced that the study was unmarred by accidental errors or deliberate falsification. And my criterion for the last was a lot stricter than that which could be proven as academic dishonesty for assessment purposes. Mere suspicion was enough. The few that I did choose to write up, however, were almost all accepted by the journals and made useful contributions. There were also many unpublished projects that helped me with development of ideas.

Of the published ones, I will mention a couple. One was published very early in my career and is still cited to the present day. This was the paper of Evans and Lynch (1973), which was the first to show 'matching bias' in Wason's famous selection task (Wason, 1966) in the psychology of reasoning – a finding that created considerable interest at the time and has remained a phenomenon of interest in the field to the current day. In fact, I have published on matching bias 40 years on from the Evans and Lynch paper (Thompson, Evans, & Campbell, 2013). This paper had a substantial and positive effect on my early career.

The second example is much more recent: Evans and Curtis-Holmes (2005). This paper showed that belief bias in reasoning is markedly increased when participants are required to respond very quickly, without time for reflection. The finding has theoretical importance and has attracted quite large numbers of citations for what is a short experimental paper. These examples illustrate two things: (a) short papers reporting single experiments can – *occasionally* – make an important contribution and become regularly cited in the literature and (b) such papers may result from supervision of an undergraduate student project. On the other hand, you could also argue that two such success stories in a 40 year career is a pretty low hit rate.

Conclusions

Most academic researchers are also teachers. It is important to do both well and find the right balance between the two activities, which inevitably compete for time and attention. While it is true that teaching, along with other university duties, consumes a lot of time and effort, it also complements research activity in a number of ways. Both involve scholarship, for example, and while teaching fields will be broader than research fields, a breadth of knowledge is of great value to a researcher. You will get ideas from your teaching that will feed into your research and vice versa. Teaching also involves interacting with students, whose questions and ideas will stimulate your thinking. In short, a positive (and efficient) approach to teaching will benefit your research work as well as your students.

KEY POINTS

- Most academic researchers will be required to spend a significant amount of their time on undergraduate teaching
- Teaching is important and must be delivered well, or else both students and the career of the individual will suffer
- Researchers with major published output are sometimes accused of neglecting teaching but there is no necessary reason why this should be true. Teaching and research are complementary activities that should benefit each other
- Linking the two activities is scholarship: the reading and understanding of literature. Teaching forces one to read much more widely, as it generally has a broader focus than research. But such reading can then engender new ideas or lead to discovery of connections between different topics
- It is a good idea to incorporate some of your own research studies into teaching: students like this and it provides an opportunity to give them insights into the process of research. Care must be taken to be impartial in grading and to instruct students to treat your work as critically as any other author
- Undergraduate research projects that you supervise can help your research as pilot studies for possible projects. These can occasionally be published as short papers provided they are of the right standard and you are convinced of the integrity of the student and the data

---

3   Designing empirical studies

Psychology is an empirical science, which means that it involves systematic observations – the collection of data. There are a multiplicity of methods available to the psychological researcher. In particular, there are quantitative methods that are primarily experimental and correlational (or psychometric) in approach and a whole host of observational and qualitative methods. I believe that the experimental method is the cornerstone of our science, but I have also used a range of other methods in my own research, such as correlational and regression methods and qualitative techniques such as diary studies and verbal protocol analysis. The precise methods chosen should always depend upon the research question that is being addressed: you need the right tools for the job.

I sometimes ask my first year tutorial students to offer a definition of psychology. They typically hazard an answer such as 'the scientific study of human behaviour'. I say fine, but how does this definition exclude a host of other disciplines such as sociology, social anthropology, economics, human geography or even history or English literature? All of these subjects (and more) involve systematic study of human behaviour. Is psychology the only one of these that is 'scientific'? If so, what do they mean by the word 'science'? I find that many students are stumped by this and have great difficulty in finding a definition of psychology that is distinctive.

A superficial try follows: the traditional method of science is the experiment, and psychology is the only one of these disciplines to use the experimental method. This cannot be right. First, it would be arrogant in the extreme to dismiss the methodologies of all these other disciplines as unscientific simply because they are not experimental (how do you do experimental astronomy, for example?). Worse, we would have to disqualify large portions of psychological research that also do not use experimental methods and often use techniques common to social sciences. Nevertheless, the fact that psychology *can* and frequently does employ the experimental method is a clue to what makes it the distinctive study of human behaviour that it is.

Let us reduce the problem thus. Psychology is a discipline that asks psychological questions. So what is a psychological question? Imagine that a psychologist and a set of social researchers of various kinds are observing some people having a conversation. The psychologist is the only one who might not be interested in *what* people are saying. The psychologist is likely to want to know things such as:

(a) What determines who speaks first and for how long?
(b) How do people know when one person has finished speaking and another may start?
(c) Which individual will have the most influence on the topic of conversation?
(d) How will beliefs of the individuals change as a result of the conversation?

These are psychological questions, as they all relate to the *processes* that underlie the behaviour. They do not depend on the topic of the conversation and are hence largely content independent.

The psychologist will form some hypotheses about these issues but will not be happy to report these as some conclusions to the research – far from it. Opinions formed from observations are not considered valid data in our subject. Some hypotheses will be suitable for investigation experimentally. Suppose we think that participants who are formally dressed or have middle class accents are more influential. We could use actors to introduce systematic differences between different groups and develop measurements of how these much individuals influence other participants. If we think eye-contact is important, we could run an experiment in which some groups can hear but not see each other and measure any differences in conversational patterns that result. Other hypotheses may, however, be more amenable to a correlational approach. For example, we might think that personality characteristics will affect who dominates conversations. So we could administer a battery of personality tests and correlate these with the percentage of time that individuals speak in the group (or use some more sophisticated multivariate method such as multiple regression or factor analysis).

The point here is that the questions are psychological because they address underlying mechanisms and processes. The questions that are addressed by correlational methods are equally psychological, even though the method used is less satisfactory. These methods are both quantitative in nature. Now there is also a movement towards 'qualitative' psychology, particularly strong in social psychology, in which researchers would be interested in what participants were saying. Conversational analysis, for example, would record the utterances of the participants and examine them for constructed meanings. The issue here relates to the question that I give my first year tutorial group. Are such research methods really psychological at all? Are they actually distinct from sociology, or textual analysis by students of literature? This is not a controversy to which I wish to add in this book. I will stick to discussion of what I understand to be psychological questions and methods. The latter certainly do not exclude forms of qualitative analysis.

In this chapter, I will discuss first the nature of psychological experiments, in the true sense, and the theoretical importance of addressing causal questions even when true experimental methods are not possible. I then look at research design as a large and complex decision problem involving a large space of alternative designs and an equally large set of criteria with which to evaluate the quality of those designs. I will argue that we need to employ research design *heuristics* in order to find the research designs that we need.

The psychological experiment

Why did I say earlier that experiments were the cornerstone of empirical psychology? This follows from my definition of the nature of a psychological question: to understand the processes and mechanisms that underlie human behaviour. Experimentation is the preferred method of empirical sciences because it allows causal inference. We set up two conditions. Independent variable A is set to one level in condition 1, and to another in condition 2. Dependent variable B is measurably different in the two conditions. If the experiment is well controlled, then no other variable changes its value (except randomly) between conditions 1 and 2 than A. Hence, we can infer that the change in A *caused* the change in B. Let us take 'well controlled' for granted. This is not a textbook on research methods, so I will not provide any general discussion of the correct methods to use to control an experiment. For this, I recommend the incomparable text of Christensen (2006).

The reason this is the basic and preferred method for psychology is that psychological questions are essentially causal in nature. For sure, there are many psychological questions that cannot for practical or ethical reasons be investigated experimentally, but the underlying questions are still causal. For example, if we investigate sex differences in cognition, we want to know if someone can be a better writer *because* they are female or a better chess player *because* they are male. We cannot, of course, randomly assign people to be male or female, so any study of sex differences will be quasi-experimental. Denied the simple control method of randomisation, we are obliged to match our male and female samples on any other person characteristics that might plausibly affect the task being measured. People often refer to studies of sex differences as experimental, but this is strictly incorrect.

There are many other quasi-experiments in the psychological literature that similarly go (often) unacknowledged. Comparing children of different ages on the developmental variable (or groups of adults of different ages) is quasi-experimental because the people cannot be assigned randomly to groups of different ages in these studies. There is also a common misconception that correlational methods can be turned into experimental ones by pre-testing of extreme groups. For example, suppose we predict that extraverts will have faster reaction times than introverts. If we simply measure a group of participants on an EPI and a reaction time task and correlate the two, this is obviously correlational. If we pre-test a larger group on the EPI and choose the top and bottom 25%, we can instead compare a group of extroverts and introverts on reaction times using a t-test. Some students now think that this is an experiment, since two groups are being compared. Of course, it is still correlational in inference, though not in statistical method.

Contemporary psychological research, quite rightly, operates under strict ethical codes in which protection of participants from harm is paramount. Hence, scientific and ethical objectives are often in conflict, but nowadays, ethics (normally) prevail. For example, our psychological question may be whether maternal deprivation hinders emotional and cognitive development in young children or whether damages to certain areas of the brain impairs cognitive function of a particular kind. Obviously, we cannot research these questions experimentally without exposing participants to harm. We make do by studying children who happen (through no fault of ours) to have been maternally deprived, or patients who have experienced traumatic brain damage. I say 'make do' from the scientific perspective, as the question addressed is still causal, even though the method adopted compromises our ability to make clear causal inferences.

Something that fascinates me about the psychological experiment – and which I rarely see discussed – is its implicit determinism. Changing environment variable A causes a change in behaviour B. That is our method and it works over and over again. Such causal influences are demonstrated and published in many thousands of psychological experiments every year. What implications does this have for those who believe that human beings have 'free will'? Of course, you may point out that psychological experiments at best demonstrate only probabilistic determinism. We almost always rely on statistical analysis of trends across groups of individuals. It is rarely the case that we can predict with high confidence what a given individual will do in an experiment, although I know of cases that are close in my own research field (for example, cognitive illusions that the great majority of people are prey to). However, if the *propensity* to behave in a certain way can be reliably and causally influenced by systematic manipulation of the environment, this does point to a deterministic account of human behaviour. We could even go so far as to say that experimental psychology is rooted in a deterministic view of human behaviour (and provides much evidence for it).

To my eyes, a psychological experiment can be a thing of great beauty. I think of it as a tool, like a surgeon's scalpel, that can be used to dissect a psychological process and lay bare its true nature. An early influence on me was Neisser's experimental study of visual search (Neisser, 1967). In a series of elegant experiments, participants were asked to search through columns of apparently random letters (comprised of rows of a dozen or so letters) for specified targets. As soon as the target was detected, they pressed a key and stopped a timer. As you would expect, the more rows that needed to be inspected prior to the target, the longer the response time.

Having established the method, Neisser used it to ask psychological questions of theoretical interest. Does pattern recognition depend upon features? If so, we might expect that finding a letter of similar features to the background is more difficult than when the background letters differ. Suppose we look for a 'T' against distracter letters that are angular – A, Y, K, for example – or against letters that are rounded, such as B, P, R. The experiments were run and the former task took longer than the latter. Can targets be searched for in parallel? Neisser observed that news clip agencies (at this time, newspaper text was not available on computers) employed people to search papers for articles for their clients, so as to extract clippings. The odd thing was that trained readers could find articles for several hundred clients on one scan. So could participants in the experiments search, say, for a B or P or R as fast as they could look for one letter? With training, the experiments showed us, the answer was yes. Hence, parallel processing was a feature of the visual search process.

I am not sure why I liked these experiments so much. I was not particularly interested in the topic area (pattern recognition). What fascinated me was the way in which the hypotheses were laid out and then tested so clearly and decisively in the experiments. I found the results of these experiments so *convincing*. They brooked of no alternative explanations. This was clearly psychological science and I think that is what I found so exciting. These experiments also seemed to close out all the issues. There seemed to be no loose ends. This last feature is rare, of course, in any experimental research programme.

Since reading Neisser's work, I have spent more than 40 years designing and conducting dozens of experiments and reading reports of thousands in the psychological literature. One comes across the odd stunning experiment that seems to provide decisive evidence between two rival hypotheses. Much more often, one's own experiments or those of other authors leave one at least partially unsatisfied, with only an incomplete understanding of the issues addressed. Experiments on their own do not do the trick: they need to be linked to hypotheses and theoretical issues with clear analytical thinking. There is also a good deal of luck involved. Well designed and motivated experiments sometimes yield unclear and uninformative findings. Sloppy research occasionally stumbles on something important.

In 1969, I attended the International Congress of Psychology, which was held at UCL that year. One of the speakers was Ulric Neisser, so I hurried to hear my hero of the time. To my dismay and astonishment, he presented a picture of depression and disillusionment about experimental psychology. The critique of experimental psychology was essentially that which he later presented in his book *Cognition and Reality* (Neisser, 1976). While equally impressive in its way, the contrast of this book with *Cognitive Psychology*, published just nine years earlier, was truly remarkable. In the meantime, Neisser had come to believe that experiments on vision used a snapshot approach that made them totally unrepresentative of real-life vision that was a dynamic process, extended over time and totally interactive with action. A highly memorable image from the 1976 book is a picture that looked something like this:

Stimulus → processing → more processing → still more processing →
  Response

This represented his *reductio ad absurdum* of the cognitive psychology of the time (just the kind that he had famously highlighted in his earlier book). He was quite right in this and, as usual, ahead of his time. At that time, linear stage models were all the rage. In my own field of reasoning, for example, cognitive models proposing a fixed sequence of processing stages were dominating big journal publication and continued to do so for some years (Clark, 1969; Johnson-Laird & Wason, 1970; Sternberg, 1980). Subsequently, psychologists devised alternative models such as those based on connectionist principles (Schneider & Shiffrin, 1977; Shiffrin & Schneider, 1977) to overcome the limitations of the linear processing approach. Later, the philosopher Daniel Dennett (1991) was to lambaste the linear approach by his parody of the Cartesian Theatre – a fixed point of consciousness in which supposedly incoming perceptual processes end and outgoing action processes begin.

The moral here is that experiments are a tool we can use to advance theoretical knowledge. They are only useful when hypotheses are carefully constructed and results intelligently interpreted. There are many soundly designed experiments to be found in the literature that advance knowledge hardly at all, as they lack an adequate theoretical context. What Neisser had really discovered was a limitation in the cognitive theory of the time. However, I believe his degree of disillusion with experimental psychology went much too far. Psychological experiments are extraordinarily useful in providing clear answers to certain psychological questions, as his own work on visual search still shows. They are not adequate or appropriate for *all* psychological questions that we might ask, however. Perhaps his most important insight was that psychological questions should select the methodology and not the other way around.

Research design as a choice problem

In the opening chapter, I suggested that conventional research methods texts tell you more about what you should not do than what you should do. That is to say a correctly designed experiment (for example) will be free of confounding variables. Of course, this does not tell you *which* experiment you should be doing. This depends upon the strategy of research, about which methods texts are largely silent.

Problem spaces and heuristics

I believe it is useful to think about the problem of research design using the language that psychologists have devised to discuss problem solving and decision making. Newell and Simon (1972) taught us to conceive of problem solving as a search in a problem space. The potential solutions are all neatly arranged in this space. The problem is to find the right one. Suppose, for example, we are trying to solve a six letter anagram in a crossword puzzle. There is a finite problem space comprised of the 6! = 720 possible arrangements of the six letters. Problem solving requires methods that are effective in searching problem spaces, especially large ones.

The problem space for finding an anagram of MASTER is illustrated in Figure 3.1. Of course, I could not show the whole space, as it is far too large. Starting with the letter M, any of the five remaining letters, ASTER, can form the next letter of a possible word. If the next letter is, say, T, then any of ASER can become

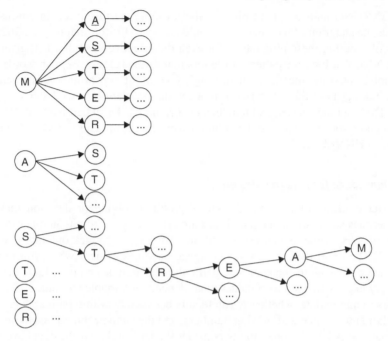

Figure 3.1 Problem space for anagram of MASTER

the third letter and so on. If we drew the whole tree, there would be 720 branches and the sequence of nodes on each branch spells out a potential solution. One of these branches spells STREAM, a valid word that passes the test for a possible solution. Of course, there might be another, e.g. TAMERS.

Now the important question is how one might go about searching a problem space such as this to find the solution. One method would be to perform a serial search of all 720 letter strings, comparing each with a dictionary (or one's intuitive knowledge of language) to determine whether it is a word. Any words found then produce a much smaller search space. If several words emerge, then the meaning of each word must be checked for its fit to the clue. This exhaustive method sounds quite efficient until you reflect on the potential search space for larger anagrams. It is not unusual for crosswords to set anagrams as long as 15 letters in length. 15! is a seriously big number. There are actually about 1.31 times 10^{12} permutations of 15 letters or, if you prefer, 1,310 billion sequences. An exhaustive search of these is hardly a feasible proposition for a modern computer, let alone a person. And yet, crossword puzzlers can solve these, sometimes even when no clues across have provided letters to reduce the problem space.

As Newell and Simon pointed out, search of large problem spaces requires *heuristics* – intelligent search methods that reduce the search space drastically in order to make the problem tractable. Suppose your crossword clue reads, 'Was Niven rung in confusion? Disturbing'. You decide this is an anagram of

NIVENRUNG and the 'disturbing' is the meaning (in this case, 'in confusion' signals the anagram). Of course, you would not try to list (or imagine) the 362,880 ways of ordering these nine letters (or even the effectively smaller list, given the three N's). You look for patterns of letters that fit the rules of English words. For example, given the meaning 'disturbing', you might expect the word to end in ING. Taking these three letters out reduces the problem to a six letter anagram plus 'ING', a much easier problem than the original. Like all heuristics, it does not guarantee a solution because the solution word might not end in ING. (Actually, it does – UNNERVING).

Problem space for research design

Now let us think of research design as problem solving or decision making. I sometimes say to my students, 'There are effectively an infinite number of psychological experiments that you could run. The trick is to pick one that is worth doing'. The enormous size of search space becomes evident when you look at the choices needed to focus on one particular research design (see Table 3.1). At the very least, you need to consider what dependent variables to measure, which factors to manipulate, whether to run factors between or within participants, what population to sample and in what numbers, and the statistical methods to be used for analysis. A lot of people do not consider the last until after the data have been collected but I would not recommend this practice.

This rather formal set of attributes leaves out a lot of other choices, not least of which is the research question to be examined! And then, of course, there are numerous details concerning materials, equipment and procedure. In practice, a lot of the thinking about design is driven by the research question itself, so let us consider some examples:

(1) Do children imitate violence that they see on TV programmes?
(2) Does working memory capacity decline with age?
(3) How does context affect anagram solving?

Table 3.1 Main attributes of the research design problem space

Type of design	Experimental (single or multiple factors, correlational, etc)
Dependent variables(s)	What will be measured and how?
Independent variables	In experimental designs, which factors will be manipulated?
Between or with participants	In experimental designs with one or more factors, each may be employed between or with participants.
Population	What population will be sampled as participants for the study?
Sample size	How many participants will be required?
Statistical methods	What kind of statistics should be used for the study?

Never mind that there are plenty of published studies addressing these questions (none of which will be discussed here). Just imagine you were set the task of investigating each of these with minimal knowledge of previous work. Research design always starts with the research question and the purpose of the study. Research designs also have to deal with specifics at a very early stage, often leading to refinement of the research question.

With Question 1, are we interested in immediate and close repetition of violent acts seen in cartoons and other forms of television, or more general long-term aggressive behaviour? Our study must address one or the other. If we are interested in immediate imitation, then an experimental study is indicated. Perhaps we could show groups of children cartoons in the laboratory, one group with and another without violent acts. Then the children could be allowed to play together with covert observation recording any aggressive actions. Long-term effects are less amenable to an experimental approach and other types of design might be considered. School children or their parents might be surveyed on television watching habits at intervals, with other measures recorded (e.g. by teachers' observations) on aggressive behaviour. Of course, this is now a correlational design and it could be that aggressive children seek out violent programmes. But note how radically different these two designs are as a result of making the research question specific enough to study.

If we look at the attributes of the research design space in Table 3.1, we can see that in the experimental version, there are numerous choices available for both independent and dependent variables. A between participant design, on the other hand, is evidently required and would require little thought to select. The population studied might give one pause for thought, as different schools have greatly differing socio-economic groups according to their location. Sample size will probably be limited by practical and resources consideration, and statistical methods are straightforward in this case. So, as is often the case, the initially large research design space is reduced radically just by thinking intelligently about the research question.

With Question 2, an experimental design is not an option because we cannot randomly assign people to be of different ages. The normal method here would be to compare different age groups' performance on appropriate tasks using analysis of variance or some other statistical method appropriate to an experimental design. But of course it is quasi-experimental, as all developmental and aging studies are, and we would have to try to match the different groups on a number of person variables. In principle, a longitudinal study would be better, but hardly practical if we want to compare age groups from adolescence to old age! Few such studies are conducted these days, anyway, with the pressure to publish at regular intervals.

For Question 3, an experimental design chooses itself, unless you want to conduct an observational study of crossword puzzlers! In this case, the design space is really large, however. The dependent variables are pretty obvious – speed and accuracy of anagram solving. But there are a number of different possible independent variables. We could provide a contextual cue, such as semantic category within which the solution word may or may not lie. This could be done explicitly, indirectly or even subliminally. Or we could use a priming approach, so that, say, three successive solution words come from the same category (e.g. animal names) with the fourth either congruent or incongruent. Between or within participant designs could be employed; problems could be arranged in randomized blocks, fully randomized orders, Latin squares and so on. So this question, although easily examined experimentally, involves a lot of choices. How one would make them, I will return to later in the chapter in my discussion of design heuristics.

Evaluation space for research designs

The Newell and Simon model is that problem solving requires exploring the problem space, identifying candidate solutions and then testing them for adequacy. With anagram solving, for example, any pronounceable string of letters found would have to be checked to see if it is a correctly spelled English word, with proper nouns normally excluded. In the case of research design, there are multiple evaluators for each candidate research design considered (see Table 3.2). Since this is also multidimensional, we can also think of it as an n-dimensional space. Each candidate research design is a point within the research space that defines its attributes. But each candidate also lies at a particular point within the evaluation space that determines its value. You might try imagining each space with only three dimensions, to simplify the task!

If we return to the chess analogy with which I started the book, then design space is like the set of legal chess moves that can be made in a given position. But just knowing what these are is useless, without an ability to evaluate the quality of each move so as to choose ones that are good rather than bad for your prospects of winning the game. Evaluating chess moves is a tricky business that the artificial intelligence programmers took many years to master (if indeed they have).

Table 3.2 Main evaluation criteria for research designs

Internal validity	*Are there any confounding variables for an experimental manipulation?*
External validity	Does the study measure what it purports to measure? Will it address the theoretical or applied question of interest?
Statistical power	Will the power be sufficient for the effect sizes of interest?
Resources	Can the study be run within financial, practical and time constraints? Are necessary facilities available?
Ethics	Does the research comply with ethical requirements?
Practical considerations	Are suitable participants available in sufficient numbers? Does the research assistant have the necessary experience? (And so on.)
Scientific merit	Is the work original and important enough to merit the resources expended on it?
Publication potential	Is the study likely to achieve publication in a journal of sufficient quality?

Evaluating research designs is also quite tricky because of the large number of aspects to be considered. In the study of decision making, this is what is known as multi-attribute decision making. The theory there is that options should be evaluated on each attribute and then given some kind of aggregate value – technically, 'utility' – for comparison with other choices. But the aggregation process is not straightforward, as some attributes are more important than others and should be given more weighting. Table 3.2 shows the main attributes for research design evaluation and choice but not their relative importance.

You will need to become expert in the evaluation of research designs. These may be your own designs that you have come up with and you are trying to decide whether they are worth running. At a later stage, you will be evaluating designs of your research students or those of other researchers whose papers or grant applications you are sent for review. As experience with research progresses, you will become aware that certain criteria regularly conflict or trade off with each other when evaluating research design.

This is a book about research strategy, not research methods. However, some technical aspects of research designs are relevant at this point because good strategy requires good choices. You will find some but not all of the attributes of Table 3.2 discussed in books on research methodology but the focus here is on the relative importance of each, the ways in which they can trade off with each other and the process of decision making that they involve.

Let us start with internal validity. This is one of two absolute considerations (the other is ethics). A design has to be internally valid (and ethically sound) or it is a non-starter. It is internally valid if it is free from confounding variables, although it may be hard in practice to identify all of these. If valid, A causes B in the context of the experiment and no other explanations are (reasonably) plausible. Internal validity regularly conflicts with external validity. Contrary to the belief of most psychology undergraduates, the external validity of research designs is a much broader question than simply the representativeness of the populations sampled. Quite simply, it is the extent to which the experiment serves the purpose (theoretical or applied) for which it was devised. In psychometric studies, the meaning of the term validity is precisely this. You may have a reliable psychometric test for aggression, but does it actually measure what it purports to measure? For some odd reason, psychologists give a lot less thought to this question when choosing dependent variables in experimental designs.

The internal/external validity conflict frequently arises because the most controlled studies tend also to be the most artificial. Essentially, there are three ways that you can control for the effects of extraneous variables on your independent variables. First, you can balance (or counterbalance) the effects of potential confounding variables. For example, if you are worried about sex differences, you can run equal numbers of males and females in each group; if you are worried about presentation order, you can run half your participants with Task A before Task B and half the other way around and so on. In general, balancing is unnecessary and causes more trouble than it is worth. For example, if you balance for sex, you have not balanced for age, extraversion, shoe size or having a row with one's partner at breakfast that morning. There are any number of organismic variables that could conceivably affect your dependent variable. In a true experimental design, they can all be controlled at a swoop by random allocation of participants to groups (this is exactly the kind of variation that the statistical tests allow for).

Aside from balancing or randomisation, you can control variables by holding them constant. In real studies, it will always be the case that a number of variables will be held at a fixed level. This is where the internal/external validity conflict comes in. Laboratory experiments are well controlled precisely because so much is held constant. Normally, each participant experiences the same experimenter, laboratory, computer screen (colour of paper or whatever) and so on. However, the external validity of the design is the extent to which your findings apply generally when none of these constants is likely to be present. If you start the other way around and choose a naturalistic setting or research method to maximise external validity, then you end up with the opposite problem: it may be difficult to show that your design is well controlled.

Another common trade-off lies between power and cost/practical consider-ations. You can always safely increase statistical power by increasing your sample size, but this is often impractical because of budgetary constraints or – just as often – availability of suitable participants. In the aggression study, for example, you are constrained by the availability of co-operating pre-school groups in your city. Yet another conflict that we have already touched upon is between ethics and internal validity. A true experimental design – by far the best way to ensure causal inferences can be made – is often not possible because of ethical constraints.

In my experience, the criteria that are most frequently compromised are exter-nal validity and statistical power. The design simply *must* be internally valid, ethi-cal and within practical and budgetary constraints. External validity is a vaguer and less easily defined concept and one that design heuristics (see below) may ignore. We would always like more statistical power (or so we think, but see Chapter 9) but we cannot always afford it or get it due to practical constraints, so we settle for what we can get.

Scientific merit is one of the hardest aspects to define and yet absolutely criti-cal. There is no point in running the research if we cannot secure our thesis or publication in the desired journal. As I pointed out in the Introduction, scientific and career motives often come to much the same thing. What is good science – as judged by one's peers – is often also what advances the careers of the scientists. Learning to judge what is good science – not simply in a methodological sense, but in the sense of developing knowledge that is interesting and important – is absolutely critical. This involves good scholarship (Chapter 1) and theory develop-ment (Chapter 5) and is also where the support of strong and experienced research groups can be of invaluable benefit to young researchers (Chapter 6).

Research design heuristics and strategies

As with all complex real-world tasks, and all decisions involving large problem spaces, we need to 'satisfice' rather than optimise. We cannot search for the optimal design because it would take far too long and, in any case, be doomed to failure. There are too many options for us to begin to consider them all, so the best we can do is choose one that is good enough for our purposes. This is the *satisficing* principle proposed by Herb Simon as part of his theory of bounded rationality (Simon, 1982). If we generated our designs at random, it might still take a very long time to find one that was good enough. What we need, therefore, are heuristics that will enable us to home in quickly on designs that are at least likely to be close to satisfactory.

While discussing the origin of ideas in Chapter 1, I pointed out that established researchers often conduct studies that follow fairly directly from their own previous research or from research reported in the journals. A lot of design heuristics are consequently based on similarity to previous or common practice in the field of study. Consider sample size, for example. How many participants provide enough statistical power for the issue under study? People often choose a number based on what is typical in published experiments. This is a pretty good heuristic. Researchers will have discovered by trial and error what level of power is needed to show effects of interest in this area. Also, editors and referees who have found those sample sizes adequate in judging whether to publish the previous papers are likely to apply the same criteria to yours.

Familiarity breeds other heuristics. For example, what population of participants should you study? Psychology undergraduate students are the most easily available. Will you be criticised for taking this easy option? Probably not, if that is what most other researchers in the field do. You might make many other choices in this way, including choice of dependent variables, methods of experimental control, choice of statistical analysis and so on. This is all very tempting as it reduces the search space in a quick and fairly safe manner. However, there are drawbacks in too much reliance on previous practice. First of all, it encourages what the Gestalt psychologists called 'blind' thinking (Wertheimer, 1961). If you always choose your sample sizes based on past practice, for example, then you are not giving any real thought to the power characteristics of your experiment. Also, just because everyone else seems to use one method for controlling variables or measuring performance that does not mean that it is necessarily the best way to do things. Some research fields actually develop bad practice in design or statistical methods that becomes calcified in their culture.

It is a commonplace observation that psychological research has a tendency to become paradigm-bound. An example, with which I am all too familiar, is the use of Wason's four-card selection task to study reasoning. Wason was exceptionally creative and relied hardly at all on the conventions and methods of others. He devised several tasks for studying reasoning of which the most famous are the selection task (Wason, 1966), the 2 4 6 problem (Wason, 1960; see Evans, 2014 for recent review and discussion) and the THOG problem (Wason & Brooks, 1979). Wason himself (Wason, 1995) was amazed when the selection task turned into a major paradigm in its own right. Hundreds of experiments have been reported in the literature using this task. I hear that the editor of an international journal has recently resolved to publish no more of them, possibly under pressure from editorial board members who believe it has become a self-serving industry. I remember journal editors 30 years ago making similar declarations to little effect. In the previous edition of this book, I boasted that I had published no papers on the task since 1996 but that record has now ended with two recent papers (Evans & Ball, 2010; Thompson et al., 2013) that somehow seemed necessary to write. These studies actually used the selection task as a tool to investigate broader theoretical questions. Overall, however, continued attachment to this task – and the broader deduction paradigm to which it belongs – has been somewhat damaging, as it has limited people's ideas about the nature of reasoning processes (Evans, 2002). In general, it is not at all uncommon for research fields to develop bad habits and sub-optimal practices that are endlessly copied. However, it is not so easy for young researchers to avoid the conventions of their chosen research fields either. I am just suggesting that you do not assume that what is habitual is necessarily the best way of doing things.

Rather more intelligent heuristics may specify default decisions that need to be scrutinised. Take the case of whether to use between or within participant designs. In general, within designs are more cost-effective because you can get the same amount of statistical power from running fewer participants. Because you need fewer participants, they are also easier to recruit and run – generally speaking, it is easier to recruit 30 participants for a one hour experiment than 60 for a half-hour experiment. (With online studies, however, shorter tasks are preferable and participant numbers can be recruited more easily for between participant designs.) By default, I normally adopt a within participant design. However, I am well aware that there are experiments for which this is an inappropriate strategy – for example, where you might get asymmetric carry-over effects, which would apply to the imitation of aggression study discussed above. So the heuristic is really 'use a within participant design unless there is a good reason not to do so', making sure that some thought is given to the condition. I have other default heuristics like this. For example, I use undergraduate student participants (cheap and available) unless there is a good reason not to do so. However, note that web based studies have now made it easier to recruit from a broader population.

Research strategies embody styles that distinguish individual researchers. One of the greatest experimental psychologists of our time, Nobel Laureate Danny Kahneman, once said to me, 'I am a main effect psychologist'. As a designer of more than a few multi-factorial experiments myself, this statement gave me considerable pause for thought. I guess what Kahneman was saying was that he preferred the elegance and clarity of simplicity in research design. Those familiar with his papers will know that the same qualities are reflected in his thinking and writing. Peter Wason also despised complex research designs and any research that relied on intricate statistical analysis to tell its story. Most of the results that stick in one's mind over the years are indeed main effects of such clarity and reliability that statistical analysis was scarcely necessary. How many researchers are remembered for their striking demonstration of a three-way interaction significant at the 5% level? I am not saying that you should shy away from advanced data analysis if you have an aptitude and liking for it. Techniques and tools continue to be developed – such as meta-analysis and structural equation modelling – that can provide important insights into scientific questions. However, I would recommend that you avoid complexity for its own sake. Even if your methodology is complex, you will need to deliver a clear and simple message when you publish your work (see Chapter 7).

A decision that any psychological researcher must make is whether to engage in research that is ethically sensitive. I personally dislike the use of deception in psychological research. I have only once published a paper in which the participants were intentionally misled (Evans & Wason, 1976). The research showed something very interesting, but even after all this time, I still feel uncomfortable about it. Researchers in cognition and social cognition have progressively adapted their methodology to avoid most uses of deception. You can give people scenarios, ask them to imagine roles and behave 'as if' they were in a particular situation with a decision to make. People seem able to do this, so that setting up situations for real (i.e. by deception) can apparently be avoided. However, there are good reasons for social psychologists to take issue with this approach. What people say they would do in a hypothetical described situation might be very different from what they actually do, as in Milgram's famous and frightening experiments on obedience. Deception aside, there are topics that psychologists study where ethical problems are endemic. For example, your research may require the manipulation of emotion, mood or motivational state. In such fields, you will spend much time and effort trying to convince ethical committees (and yourself) that your research will not harm your participants.

Research styles and strategies can be low or high risk. If you adopt the strategy of carefully reading the current literature and designing experiments that fill in the gaps left by other researchers, then you will most likely be published provided the methodology is sound and the write-up competent. Such research is necessary and an important aspect of what Kuhn (1970) calls 'normal science', where researchers are working within an agreed paradigm. A higher risk and reward strategy is to aim for innovation in theory or empirical study. This approach is much harder because you set aside many of the heuristics based upon following conventional practices. You take nothing for granted. Even though no-one has heard of you yet, you take the view that your opinion is as good as any of the published greats in the literature. No matter how big the name or the journal, if you see something wrong with the theory or the experiments, then it is wrong and you can try to establish this with your own research. If you are dissatisfied with the conventional methodology, then you challenge it and propose an alternative.

The high-risk approach is an ambitious one because if it succeeds, then in due course, you will be known, not simply as a solid and reliable researcher, but as an innovator and leader in the field. However, it is less likely to succeed unless you do, in fact, have some exceptional talent. It is also worth a note of warning about the way that some young researchers, eager to establish their reputations, can come across to their seniors. I have seen cases of individuals who have no exceptional quality of intellect and no important insights to offer early in their careers taking an aggressive approach in their early papers, dismissing the work of established authors. The consequences are not good: such researchers struggle to get papers accepted by the journals or grant applications funded. Such papers, if they are published, tend to irritate the established names in the field. Never overly fond of criticism, such established researchers are particularly loath to take it from relatively unknown authors who show little depth of understanding. All research fields have an established hierarchy of editors, authors and referees that can powerfully influence the career prospects of young researchers.

My advice on this is as follows. Do not overly respect established authors and research practices when designing and interpreting your studies. You should certainly not feel in awe of the big names. Take the view that your ideas are as good as anyone's until proven otherwise but show respect for other authors at the same time. When you present your work at conferences or in journal articles, take care to come across as confident but not arrogant; as an independent thinker but one who has nevertheless carefully studied and assessed the established work. If your work is innovative, thoughtful and carefully argued, then that is the way it will come across and your reputation will soon grow. There is also no harm in establishing yourself with some solid contributions in early career. You can still show qualities of imagination in design and interpretation of your findings that will catch attention and interest.

Conclusions

Research design involves a lot of choices. In this chapter, I have invited the reader to view the process as problem solving and to adopt the Newell and Simon approach of heuristic search of problem spaces. The problem space for research designs is very large and each design has to be evaluated with regard to multiple attributes (see Tables 3.1 and 3.2). It encapsulates the chess problem: how to find which of the many available choices are actually worth taking.

This chapter cannot be read in isolation, however, as everything else that you do: reading and scholarship, theory development, collaboration, etc., feeds into the actual designs you choose. You start – *always* – with one or more research questions. Asking the right questions is the really hard part that requires creative thinking (see Chapter 5). Choosing the right design to address them is important and can be complex, but is largely a matter of technique. Heuristics used to reduce the search can be explicit and conscious but all too easily become automated with experience. This is not necessarily a good thing, as habitual, thoughtless choices are not necessarily the best ones. Sometimes, breakthroughs occur when people re-think their methods and design choices in a radical manner.

KEY POINTS

- This chapter deals with the *strategy*, not the methods, of research design. In particular, there are many choices to be made about the design of studies even once the research question is established

- Research designs differ on many features (experimental or correlational, between or within participants and so on), so design can be thought of as making a choice within a multidimensional design space (Table 3.1)
- Each candidate design can be evaluated on a number of different aspects (e.g. internal validity, power, cost, ethics) and so can be seen as lying in a multidimensional evaluation space (Table 3.2)
- Given the complexity of choices and evaluations, the search space must be reduced by the use of *heuristics*: intelligent methods that contain the space of possibilities to something more manageable
- The most common heuristics are based on familiarity. What is typical practice in the literature may point to choice of paradigm, testing conditions, sample size and many other features of the design. Too much reliance on common practice may limit creativity and innovation, however
- Other heuristics may make use of default choices. It may be easier and more cost effective, for example, to use within-participant designs and select undergraduate psychology students as the participants. Such defaults cannot be applied blindly and must always be checked for suitability
- Good research design choices require that questions of scientific interest are being addressed. This depends in turn on scholarship, creativity and theory development – topics covered in other chapters

4 Research funding

Arguably the worst part of the academic researcher's life is the continuing need to generate research funding. Unless you are lucky enough to be part of a centre with long-term programmatic research funding (which is a rare exception), then you will need continually to worry about generating research funding from sources external to your university. You may be well funded at a given point, with a group of research staff and students around you, but in two years' time, they could all disappear (easily). It does not matter how good your reputation and track record is. Most of your research will need to be funded on a project by project basis. If you are doing theoretical research that involves application to public bodies, then your applications will be peer reviewed in a highly competitive system that rejects the majority of grant applications that are deemed, in principle, to be fundable.

Research funding is a very important topic but also the most difficult to write about in a book like this, which aims to have cross-national appeal. The reason is that funding systems vary a great deal between different countries. In the UK, the system I know best, research in universities has for many years been 'dual funded'. Funding is supplied directly by government to establish the infrastructure for research, using the peer reviewed Research Excellence Framework (REF, previously known as the Research Assessment Exercise), where money is allocated on the basis of the recent research record of the department concerned. Where funding is large, good laboratories and facilities can be founded and more academic staff or faculty employed. This enables lighter teaching loads and more time to be spent on research work. Even in this system, however, academics are still expected to apply for research grants from external bodies to fund particular research projects, including a number of publically funded research councils. Career progression can depend critically on the success one achieves in this highly competitive enterprise.

By contrast, some countries focus more on funding researchers than individual research projects. For example, in France, researchers may be funded by central body CNRS, either in centres or on placement within university departments, with a lot of freedom to choose what research is conducted. Similarly, elite researchers are gathered into the prestigious Max Planck institutes in Germany. In Canada, research funding is provided to university academics on a more or less universal basis, but with relatively small amounts spent on projects than, say, in the UK.

My Canadian colleagues are always amazed by the size of UK research grants and unconvinced that the research needs to cost that much! This is partly a matter of accountancy practices, as in the UK, the full economic costs (including, for example, the salary for the hours spent on the research work) have to be funded. The Canadian system also seems to make a lot more use of students to run research projects, whereas in the UK, projects usually employ more expensively salaried research assistants and research fellows.

In the United States, as I understand it, there is similar pressure to fund research projects via competitive grants as in the UK, with governmental funding bodies (such as the National Science Foundation, Washington) as well as other government agencies, private corporations and charities able to award grants. The pressures and processes involved in obtaining funding seem to be quite similar between the UK and the USA. The management of grants once awarded is somewhat different, though. In the UK, spending has to be very closely accounted for with respect to the particular research specified, as well as the various headings and subheadings for equipment, travel, staff costs and so on. My impression is that my American colleagues have a great deal more latitude in how they spend their grants once they are awarded.

In spite of these cross-national differences, I believe I can offer some quite general advice about how to obtain, manage and report on grants for particular research projects.

How to get funded

The research funding systems as I know it is inefficient and frustrating. This is because you will have to spend a large amount of time writing applications that may not be funded. It is a very competitive as well as a somewhat political process. The ideas may be good, the research sound and well-costed but something else gets priority when the funding committee meets. It is common to get a response that says, in effect, that we thought your research project was suitable for funding but could not find the money for it. With journal articles (see Chapter 7), there is usually a constructive route to follow if a paper is rejected. For example, you may be able to resubmit to the same journal and certainly can do so to another, after suitable re-writing. Rejected grant applications are more difficult to deal with. You cannot normally resubmit to the same funding body with small changes. You can submit elsewhere *if* there is an alternative funder suitable for this research work. But this will usually involve a lot of work, as the format of applications required can vary considerably between different funding bodies.

My personal strike rate over my career was around 60–70%, which may not sound very high but is well above the base rate. I had runs where several applications were funded in succession, but also where several were rejected. Statistically, this is bound to happen, but the emotional impact on the individual can be very strong. Repeated rejection, particularly in early career, can be very demoralizing. If it happens to an experienced researcher, then it is also very worrying, as you wonder if you are losing your touch or losing favour with your colleagues. In truth, there are just too many good applications chasing too little money. But you *have* to take this on. Research funding is considered so important in universities and has such impact on your career that you cannot simply opt out. There are ways of running research unfunded, which I discuss later, but this is a last recourse.

Given this situation, there are both quantitative and qualitative strategies for getting funded. I know researchers who use both approaches. The former crank out lots of applications, expecting the majority to be rejected. They take the view that there is a lot of luck in the system, depending upon which referees you get and who is feeling argumentative on the day that the funding committee meets. Throw enough mud against the wall and some will stick. I know researchers who take a similar attitude to publication in good journals, too. The qualitative approach involves trying to write applications of exceptionally good quality. You write fewer applications, taking more time over their preparation, but expect to achieve hit-rates well above the base rate. You still have to accept that some will be rejected. Researchers taking this second approach tend to have more belief both in their own ability and in the integrity and reliability of referees.

Managing a long-term academic research programme is quite tricky. You will want to get to a point where all aspects of research are going on at the same time. That is, you are writing up results of completed research projects while supervising the conduct of current projects and also preparing applications to secure funding for future projects. It is the unpredictability of the research funding system that causes the most difficulty. Say you make two or three applications per year, needing at least one to get funded to keep things going. It can happen that all of them get funded in a given year – or none at all. It is somewhat easier in large research groups where a number of collaborators can make more applications between them, smoothing out the random fluctuations to some extent. Nevertheless, you will need to develop strategies for coping with both too much and too little funding at particular times. For example, you could take advantage of a lull in funding to do more writing of papers based on completed research that you have been too busy to get to. If you are short of data, you could work on a theoretical paper, or a review or a book. And, of course, you have plenty of time for writing new grant applications!

The question of how research grant applications are framed is closely related to issues discussed in previous chapters. They will reflect your style and ambition as a researcher. However, there are certain basic rules you can follow. If you are an early career researcher without a track record of successful grants, you are unlikely to be awarded a large grant on your own. If you are fortunate enough to be in a large research group, you can team up with more experienced researchers, in whom the funding committees will have established confidence. If you need or wish to make your first applications on your own, then go for relatively small sums until you have proved that you can deliver.

There is a common mistake in how young researchers typically present grant applications. The most important section is the plan of work, where you specify the research to be carried out. Most inexperienced researchers will devote far too much of this to theoretical argument and rationale and provide insufficient specification of what they plan to do. I have seen applications where 80–90% of the space allowed for the proposal is spent on rationale. The experimental programme is presented almost as an afterthought and amounts to little more than 'we will do some experiments to test these ideas'. Sometimes, you find that your little known author is asking you to risk a substantial amount of public money on this unspecified enterprise. Such applications have almost no chance of success.

It is important to put yourself in the position of the referees and committee members. They are entrusted with distribution of a scarce and valuable resource. There is never enough money to support science. They have two main priorities in mind: (a) to support good work and (b) to ensure that there will be a return on the investment. According to the funding body and political context, they may also be under pressure to fund work that has prospects of early impact in terms of application. Established researchers have an advantage because they have shown that they can do good science, as evidenced by their publication record, and they have also managed successful funded research programmes before. People want to encourage and fund young researchers but are cautious about committing large sums. Also, they need a proposal where they can see that the research is likely to succeed. High risk research is always hard to fund, and almost impossible in early career.

The second point applies no matter how experienced the applicant. You will not get funded simply on reputation (this may happen, but much less than you might think). Hence, successful applications normally are very clear indeed about the purpose of the research and how it will be carried out. Try to use no more than 20–25% of your allotted space to describe the background and rationale, but make the objectives extremely clear. The empirical programme should then be specified in as much detail as possible, even to the point of describing the outline design, materials and procedure for studies 1 to 6 (say), specifying the size and nature of the samples of participants to be tested. Reviewers can then see that these experiments are soundly designed and do indeed address the objectives specified. This increases their confidence that the project will deliver. It also makes it much easier to judge whether the resource requested is appropriate.

I know researchers who find writing grant applications very difficult because they have the opportunistic research style. They like to run an experiment, think about its results and then come up with another experiment, repeating the process. A lot of PhD's happen this way, so it may form a model for young researchers when they move to the postdoctoral phase. For some people, this is an effective way of conducting research, but in truth, you are unlikely to get a research grant that specifies such a strategy. It just comes across as 'trust me, I will think of something as I go along'. You simply have to plan ahead a sequence of studies that can be presented in the way I describe above. But some people will say: what if the results of Experiment 1 are not as expected – will that not throw the whole sequence off?

Everyone knows that research does not go as expected and that design of later studies will have to be modified in the light of early findings. When you come to write up a report to the funding body on the completed project, you are able to explain and justify such deviations. At this stage, the funding body is more interested in what the project produced than whether it stuck rigidly to the initial plan. However, it is important to show that you still pursued the objectives of the original proposal unless there was a very strong reason for altering them. So, you might say, what is the point of specifying a research programme in such detail if it is almost certain to change when the research is run? The bottom line is that you will not get the grant unless you do! If you want a more rational reason, it is that you at least demonstrate to the reviewers that you can design studies appropriate to your objectives and cost them correctly.

Choice of projects and funding bodies

One approach to research funding – 'bottom-up' – is basically to consider what research it is that you want to do and then look for a way to fund it. That has been my own approach in the main. You need to have confidence in your own ideas and an established publication record to obtain support. From this point of view, the motivation for the research is intrinsic. Your goal is to carry out a piece of research; the funding, a means to that end. The problem with this pure approach, however, is that you make it more difficult to get funding in a number of ways. First, you are restricting the number of funding bodies you can approach – specifically, those that will fund whatever kind of work you want to do. Second, you are normally unable to benefit from various 'top-down' initiatives that might be out there.

Research funding is a very political business and not just in the sense of academic politics. For example, government ministers rarely share the academic ideals of knowledge for its own sake. Nor are they keen on trusting academics to come up with the right work on their own initiative. Depending on the funding body, there may be more or less interference with funding processes as a result. In the UK, for example, great pressure has been put on research councils in the past 20 years or so to make sure that work is relevant and applicable to the real world. This can affect the ways in which applications have to be made, especially for basically theoretical work. It can also result in funding initiatives where large pots of money are earmarked in a 'top-down' manner for particular topics or themes.

A more pragmatic approach to funding, then, is to judge the political mood and be aware of special initiatives and reserved pots of money. Then you can design the kinds of projects that the political masters deem worthy. The research is now being driven less intrinsically and more by the chances of funding. A lot of people take this approach, although I never have. My intrinsic motivation to address my own research questions was simply too strong. There is also a political aspect to research funding that I dislike and have had little to do with. It helps to be *connected*, to know the academics who sit on the funding committees and may give advanced warning of special initiatives as well as other insider advice. I have known public bodies announce special funding initiatives with deadlines so short that no-one without advanced knowledge would have a chance of preparing an application.

Some research funding is intensely political. I have particularly in mind the vast amounts of research money available from the European Union. It is frustrating that they allocate such large sums in the way they do. Not only is the great bulk of the money for top-down topics and initiatives, but there are a number of political constraints. Normally, you have to include academic and industrial partners from several different European countries, including sometimes those with little research strength and tradition that the EU wants to encourage. As if that was not bad enough, it is very hard to get funding without active lobbying of EU officials in Brussels. A number of universities employ their own lobbyists precisely for this purpose. The great bulk of funding is also aimed at projects with immediate impact and application – a problem for theoretical researchers. This is definitely one for researchers who start their search with the money rather than the project. Needless to say, I have never had an EU grant.

For some researchers, the attainment of high levels of external research funding becomes an end in itself, something encouraged by university politics as I discuss below. You do not need to confine yourself to the public bodies and charities that fund more basic kinds of research. You can also apply to pharmaceutical companies and other industrial sources for funding. In this case, the research will have to be very applied and there may be issues about publication and intellectual property rights more generally. This may also lead to opportunities for consultancy work with fees paid for expert advice. Universities have different policies in the extent to which this kind of work is encouraged and whether staff can retain some or all of the consultancy fees. I do have some experience of work of this kind. For example, at one stage, Educational Testing Services, Princeton, approached our reasoning group for help with some of the reasoning items in the Graduate Record Examination test, which led to a series of substantial contracts.

Managing and reporting on research grants

Let us suppose that you have obtained your research grant. You have your programme of work and a budget for, say, two to three years. Now you need to conduct the work, publish the findings and report back to the funding body at the end. You will have to comply with the financial regulations of both the funding body and your own institutions and there will be administrative staff in the university to help with that. Otherwise, essentially, you are on your own. But the way in which you manage the project and the report you present at the end could be critical to your career. If the research does not deliver its objectives and the funding body is not happy with the report, you could have great difficulty in obtaining further grants. A first major grant, in particular, is an awesome responsibility.

Early career researchers will at least have had the experience of their own PhD work to draw on, even though it was supervised. If you managed the time well, conducted the planned studies to time and budget and wrote up the thesis quickly and efficiently, you probably have a good idea of what you are doing. If you failed on one or more of these counts, then be sure to learn from the experience because you cannot afford to repeat the mistakes on your first research grant. I would strongly recommend that, if at all possible, you collaborate with a senior researcher with plenty of experience of managing funded research programmes. If you involve them in the application itself, then you will improve your chances of getting the grant, as it will seem less risky to the funding body. But bear in mind that it is also possible to recruit collaborators and co-authors after a research grant has been awarded. However, they will not share your responsibilities so far as the funding body is concerned.

The most important priority in managing a research grant is to complete the work you planned or its near equivalent. If the work changes direction because of unexpected findings or methodological issues, then you should be able to justify that in your final report. So long, that is, that the quantity and quality of the work is equivalent to that in the proposal. What you cannot afford is to plan three studies and run only two, or run the third at the last minute and have no results and analyses to present in your final report. Most definitely, you cannot report that studies were incomplete because you had insufficient time or money to run them. That makes you look completely incompetent, as you were the one who planned the research and specified the resources needed in your application. The sponsors can only conclude that you are incompetent in either the planning or conduct of research, or both.

It is also not a good idea to rely on extenuating circumstances to explain failures to deliver the project objectives. You may believe, with some justification, that your research was held up because a research assistant resigned half way through and you had to recruit and train a replacement. Or that it took longer to obtain ethical clearance and other forms of consent than anticipated; or that participants proved difficult to recruit at the time you needed them; or that the workshop was too busy to set up your equipment on time and so on. Putting any of this in your final report, however, will just look like excuses. As you will learn with experience, things of this kind almost always go wrong at some point in a grant and you have to build in enough contingency for them. There is an important implication here for the writing of the initial application. Do not be overambitious in the hope that the large amount of work you plan will impress the funding committee. As a general rule, if things can go wrong, they will. The basics of managing a research grant project are really quite similar to those for planning a PhD programme, which I discuss in Chapter 6. Do not be deceived by the feeling of time to spare that you might experience in the early stages. Make a timetable at the start and stick as closely to it as you can. Build in some spare time for the unanticipated things that will hold you up and so on.

Next to completing the programme, the most important thing is to publish your work and publish it well. The question here is timing: when should you be writing up papers? Ideally, you should not work for more than a year or so before submitting papers for publication. It will certainly help your final report if you can already list papers in press supported by the grant. Because papers can be rejected and resubmitted, the earlier you can start the process, the better. But this is not as important as completing the planned work on time, if it comes to a choice. Strictly speaking, writing for publication is part of the research and should be completed along with everything else. Ideally, all work would be either published, in press or submitted at the time of the final report. In practice, you will get a bit more leeway on this than you will on slow progress with the empirical work. Everyone understands that publications tend to lag behind research work. So my advice is to write work up as you go along but not at the risk of failing to complete the main programme of work.

You typically have a few months after the grant completes to write and submit the final report. This report will normally go to academic referees who will compare what you achieved with what you planned. While part of the report will deal with finances and deliverables, it will still read like a mini-journal article reporting the research itself. It will generally be shorter than a journal paper but the same principles apply (see Chapter 7). You must tell a clear story that engages the reader's interest in a satisfying way. The skill with which you can write this report may affect the final rating quite apart from the quantity and quality of the research work conducted. If your research diverged significantly from the plans, you must provide a good explanation as to why and demonstrate that good findings resulted from your decision to change direction. If you can back that up with published papers, then so much the better.

Working unfunded

The extent to which you can conduct empirical research without external funding depends on what you are trying to do. If you are a neuroscientist planning fMRI studies, you can forget it. It you work in social cognition and can conduct studies using pencil and paper vignettes, you may have a shot. You can probably get some help from undergraduate project students (see Chapter 2). You may have to prepare materials and run analyses yourself, of course, in the absence of a research assistant. Recruiting and running participants is the most time-consuming part for an experimental psychologist. One possibility is to hand out booklets during a lecture, but I found this practice, once common, was increasingly discouraged over my career. Even if it only takes a few minutes, students will complain that the time is being taken out of their teaching. For this reason, some departments may ban the practice altogether.

There is, however, a very good alternative available these days: web based research. The use of web research has become acceptable to editors of psychological journals, with initial worries about lack of control over participants and testing conditions having been apparently set aside. Of course, not all research can be administered this way but the software packages available now allow for quite a range of studies to be delivered remotely. There may be some costs in accessing the right websites to recruit participants but these are relatively minor and can probably be claimed from departmental budgets. So provided you can develop the skill to program your own web experiments, this is a way of keeping certain kinds of empirical research going without external funding. As alluded to in the previous chapter, design choices for web studies may differ from those run in the laboratory.

If you are not currently funded, however, I would recommend that you devote a significant portion of your available research time to writing grant applications.

In most psychology departments that I know, there always seem to be one or two individuals who maintain a published output of some kind without external funding, having perhaps given up on the tedious and sometimes soul-destroying process some years ago. Such individuals are generally not too popular with their academic managers, however, and certainly do not carry the status and clout of the big grant-getters. I will return to this in the final section of this chapter on the politics of research funding within universities.

Research fellowships and sabbatical leave

So far, I have discussed only funding for empirical programmes of work. However, there is a different kind of funding – one that pays for your teaching and other duties to be covered, providing you with a block of free time to think and write. In highly conceptual and theoretical subjects such as mathematics and philosophy, such personal fellowships may be the main and most appropriate form of funding. After all, you cannot employ a research assistant to do your thinking for you. It is possible to obtain externally funded fellowships of this kind, although in strongly empirical subjects like psychology, they are exceptional. The reason is quite simple: it is much cheaper for a funding body to pay the salary of a research assistant than a faculty member. Also, as one gains experience, there are a lot of routine aspects of research work that are more efficiently delegated to others.

However, it is as true in psychology as in philosophy that you cannot employ someone else to do your thinking and that one can benefit greatly from a period of time free from other duties. I was lucky enough to receive one such period of funding late in my career from the ESRC, which paid my salary for 30 months while I worked on developing the dual process theory of reasoning. This was one of the most enjoyable and productive periods of my research career. Such grants are generally only available to senior scholars who have had time to acquire sufficient knowledge and reputation to justify them. However, many universities do follow the practice of giving their academics regular periods of sabbatical leave – perhaps for a year, but more commonly, six months – where they are left free of all normal university duties. These may be awarded to junior as well as senior staff. Practices vary greatly between countries and universities; it may be automatic or discretionary, fully or partly salaried, given at frequent and infrequent intervals and so on. You may have to apply for sabbatical leave and make a case for how you will spend your time, reporting on it afterwards. In this case, it is much like a research grant, except that your own university is the funding body.

I was fortunate enough to benefit from regular periods of sabbatical leave throughout my career and cannot speak too highly of the benefits to individuals and departments of allowing this. Of course, it has to be funded in some way. In my department, we had sufficient numbers of staff to provide some teaching and administrative cover, all recognising the benefit to us all in supporting the system. Typically, leave would be for six months and the recipient might expect a higher than normal teaching load in the other half year. However, we had a strict policy of not contacting an individual on leave (even if they remained in their usual office) over regular university matters. We all understood the principal benefit of such leave – a free mind. Individuals will, however, normally need to cover research related duties such as PhD supervision or oversight of current research grants, or make arrangements for these if travelling extensively away from the home university.

How should these periods of leave best be employed? The first decision to be made is whether to travel and spend time at another university, perhaps in another country. In some universities, and in some countries, this is expected. If you are isolated from the best research in your field, it makes sense to visit somewhere that has excellence in the topic to develop links and collaborations. Generally speaking, the earlier you are in your career, the more beneficial this is likely to be. This may not always be a feasible option, however, if, for example, you have children at school or you have no financial support for travel costs. I only had one major travelling sabbatical in my career when I spent the first six months of 1982 at the University of Florida, taking my wife and four children with me. It probably did not do a great deal for my academic development, but it was certainly a vivid and lasting experience for me and my family. It also helped me understand the culture of universities and academics in the USA, especially as I taught an undergraduate course to cover the travel costs.

As I said above, the big benefit of fellowship leave is having a free mind, so different from a research day with teaching and other duties always hanging over you. This is particularly useful for scholarship activity, theoretical development and writing. I have written a number of academic books over my career and all (except this one!) benefitted from a period of research leave. Most of the books were in the form of research essays, so I was not just writing, but thinking – developing new concepts and theoretical ideas. This is so much harder to do while engaged in regular university duties than is the supervision of empirical research projects. However, I am not suggesting that my activity would be the norm. Many successful academics never write a single book in their careers and do not necessarily try to develop original theories. Even if you are highly empirically oriented, however, the free mind will help you in such matters as (a) writing papers for major journals, which requires exceptional clarity of thought and expression, (b) reading your way into new topics when you wish to change the direction of your research and (c) preparing ambitious grant applications.

Sabbaticals and other periods of research leave are a great resource. They should not be squandered but often are. What can go wrong? The general mistake is the one that I discuss for PhD students, setting out with a long period for research (see Chapter 6). That is, thinking that you have plenty of time and starting without a particular plan or structure. If you made this mistake as PhD student, it does not mean that you will not repeat it! You should always work with specific objectives and a timetable against which you can assess progress. The time to do this planning is *before* you take the leave, not during it. You do not want to waste a minute of this precious time doing stuff you could have done in advance. Nor do you want to waste time floundering around trying to think of ideas. It is criminal to waste a sabbatical: you may have to wait a number of years for another such opportunity.

University politics and research funding

I have already discussed the external politics of research grants: how their nature and chances of award are influenced by government pressures and by the degree of connectivity that individual researchers have with influential academics in other universities. Research funding is also very important within universities in determining status and favour with the senior academics running the place.

It would be nice to think that university managers see the purpose of the institutions as the pursuit and dissemination of knowledge: nice, but naïve. There are two big drivers of university policy: money and prestige. The two are linked, of course, as the most prestigious universities are also, in general, the richest. To an extent, universities can be seen as businesses that need to make a profit and reinvest it to improve the future. Certainly, they cannot afford to make heavy losses unless operating in a country where they are fully nationalized. They need to generate more income than they spend. Most university bosses are keen to see money of all kinds coming into the institution. The contribution they want and expect from individual academics is research grants: the bigger, the better.

I heard a story some years ago about a young lecturer in a leading UK psychology department who obtained his first research grant from a government research council. The grant was for about £20,000, which at that time would have paid for a research assistant, plus costs of materials and participants for a two year project. The lecturer was called in to the office of his senior and eminent Head of Department. The conversation went something like this:

Head: I see you have obtained your first research grant?
Lecturer (expecting praise):   Yes, indeed.
Head: Why is it only for £20,000?
Lecturer: Uh, well, that is all I needed.
Head: Don't be so naïve! Find a way of making your research expensive!

I am not sure if this actually happened, but it may well have done. In fact, I imagine that a number of exchanges along these lines have taken place in many universities. The RAE/REF system in the UK has made things worse by giving credit for the scale as well as number of grants awarded. So what that comes down to is this: judge the value of research by its cost.

This worries me quite a lot. Philosophy departments are an endangered species in the UK now. Admittedly, this is partly due to lack of student demand but philosophers also have the big political problem that they do not need research grants. Their research is far too cheap. By contrast, university managers love to appoint neuroscientists because their research is so expensive. It is also fashionable, so that they can obtain large grants. I have a lot of experience running reasoning research without neuroscience methods, but also some with, in recent times. I can attest that a grant with the methods included greatly increases the cost of the research and, hence, the value awarded. Whether greater knowledge and understanding is acquired as a result is debatable.

Two things bother me particularly. First of all, the value of research is quite evidently *not* its cost. I can think of a number of examples of books or theoretical papers published by psychologists that have radically changed thinking on the topic, without any expensive empirical research of any kind. On the other hand, I frequently see very expensive pieces of empirical research published that contribute little or nothing to our knowledge. The second problem is that research funding is not obviously real income to the university in a business sense because most or all of it gets spent on the research itself. The award of a grant may even end up costing a university money because of the indirect costs that department has to meet.[1]

This brings me to the second driver of universities: prestige. Or, if you prefer, academic snobbery. Larger research grant income provides a bragging right, much easier to quantify than contribution to knowledge. Moreover, because of the prestige factor, not all money is equal. In the UK, money from government research councils, for example, is generally valued a lot higher than money from industrial sources. This is because it is perceived as funding more theoretically based and, hence, *cleverer* research than that obtained from industrial sources. It is true that, under political pressures, the research councils increasingly require candidates to demonstrate 'non-academic' impact of their research work but it is still scientists and not politicians or industrialists who sit on the funding committees. Whatever the rules of the game, academic referees will tend to favour research with a strong theoretical basis. The UK REF system explicitly acknowledges this, giving more weight to traditional academic sources, in the same way that it gives more weight to articles published in prestigious journals. Elite universities want to say: we have the best academics, the best research and attract the most (good) research money.

This is an advice book, but I am not going to tell you how to respond to this political climate: I just want to make you aware of it. Personally, I think it quite ridiculous to make research more expensive that it need be: immoral, in fact, when the taxpayer is ultimately funding it. I also cannot imagine choosing to give up a field of science that fully engages my interest and moving to another on the grounds that larger research grants are easier to obtain. I have never done anything other than seek funding for the work I wanted to do at the expense it genuinely required. But your chances of promotion or to appointment to another university will, to quite a large extent, be influenced by both the number and size of your research awards.

KEY POINTS

- Obtaining research funding is a difficult, frustrating but essential part of the working life of most academic researchers
- The research funding strategy you adopt must depend on your research style and the nature of your academic ambitions

- One approach is to identify the research questions you wish to pursue and then look for a way to get this work funded. Another is to look at what funding is most available and be prepared to pursue questions compatible with funding initiatives and other political constraints. There may be a conflict between pursuit of questions of interest and pursuit of research funding
- Research grant applications must be exceptionally clear in defining research questions and methods. They should not be overly ambitious for early career researchers. Funding bodies will want to be confident that there will be a scientific return for the money spent
- Managing funded research programmes well is of great importance. It is essential to complete the work proposed and, hence, to build in a contingency for delays and problems in the original proposal. Early publication of work is also desirable and will help you to impress in your final report to the funding body
- Sabbatical leave or externally funded research fellowships provide great opportunity for theoretical development and writing with a mind clear of other duties. Such leave can also easily be wasted
- In university politics, the size and frequency of research grant awards will be of great benefit to the career of an individual, even though the actual costs of research work vary greatly between disciplines and topics

5   Developing and testing theories

This chapter is not just concerned with theories but with *theory*. Theories and models are sets of proposals, more or less formally stated, that can be tested by empirical science. You may test theories that are formulated by other authors, or modified or invented by yourself. Theory, on the other hand, refers to the knowledge and understanding that you build up about your research topics based on scholarship and hard thinking about research findings of your own and other authors. To be sure, theory, in this sense, can lead you to formulate specific theories and models for test. But it is a broader notion. Without theory, you cannot formulate research questions, write introductions to journal articles, frame research grant applications, interpret findings and so on. This is the essential raft of ideas and understanding that makes you an academic. There are technical and philosophical issues surrounding the formulation and testing of theories that I leave to Part Two of this book. I include this chapter in Part One because without theory, you cannot do any of the other things I talk about: design studies, prepare grant applications, write journal articles, etc. So how do you go about developing theory (and theories)?

In Chapter 8, I will discuss the logic of hypothesis testing and some philosophy of science. In particular, I contrast the philosophy of Karl Popper (falsificationism) with that based on the Bayesian approach. Popperians, apparently, spend their lives trying to find experiments that will refute their own theories, gladly giving them up when an odd experiment produces an inconsistent result. If scientists like these exist, I have never met one. Bayesians, on the other hand, build up beliefs in theories or lose faith in them in a gradual manner. They deal in uncertainties. Not only are they uncertain about their theories and hypotheses, holding these with a degree of belief, or subjective probability, but they are also uncertain about their experiments. Experimental findings also hold with some probability, due to uncertainty in methodology and the possibility of error – human or otherwise – in experimental design and data analysis. Scientists that I know are much more like these imaginary Bayesian creatures.

As I indicated in the Introduction, there are two quite opposed views of what science is and of what scientific knowledge consists. I again will caricature extreme positions to make my point. The Empiricist's view is that science is a collection of facts built up by careful empirical study from which theories emerge. The Theorist's view is that science consists of understanding encapsulated in well-developed theory. The role of fact collection, according to this view, is to assist in the development of the theories. Some of the philosophical issues discussed in Chapter 8 revolve around this distinction. Empiricists want to describe the world by inspection of the facts, and hence have to deal with the problem of logically invalid, inductive inferences in order to generalise from these observations. Theorists, on the other hand, attempt to construct the best (most complete, most parsimonious, most predictive) theory that is consistent with the available facts and use this theory to guide the design of future empirical research.

For an insight into the mind of a theoretical scientist, I strongly recommend the informal autobiography of the famous theoretical physicist Richard Feynman (1985). He and his colleagues used to have a saying: 'Never believe an experiment until it is proved by theory'. On occasion, he would be absolutely sure that an inexplicable experimental finding was rooted in methodological error and would instruct his experimental colleagues to find it – which they generally did. Theoretical physicists inhabit a curious and wonderful mental world in which apparently no thought is too weird to be contemplated: negative time scales, spontaneous creation of matter and anti-matter from nothingness and so on. Of course, they are not just extraordinarily imaginative but also highly rigorous and possessed of the strongest mathematical skills. To be sure, they think about empirical findings, but it is the thought itself that seems to be the main tool of their science.

Physics is more or less unique in drawing a sharp division between theoretical and experimental science. The great majority of psychological researchers, for example, design and supervise the conduct of empirical studies. However, none but those focussed on the most specific of applied issues can avoid the responsibility for the development of theory. While innovation of a major theory is a fairly rare event, the publication of empirical papers in good psychological journals requires that the research be theoretically motivated. That is to say, the research must be designed to test and advance theory, in the sense of knowledge and understanding, and the discussion that authors provide of their results (see Chapter 7) is an absolutely crucial part of the contribution.

I have talked in an earlier chapter about research styles and they are relevant here as well. Programmatic researchers who pursue more or less integrated research themes over many years – sometimes over a whole career – will build, develop and test a large body of theoretical knowledge. The most successful and revered researchers fall into this category. In psychology, Piaget and Skinner would be good examples of the most successful of programmatic researchers, even though much of their own work and that which they inspired in others was empirical in nature. There are, however, many examples of strong psychological researchers of good (but not stellar) reputation who follow a strongly theory-driven, programmatic approach to their own research careers. I know some equally strong psychological researchers, however, who follow a model more similar to that of the experimental physicists. These individuals are by no means lacking in theoretical skills, but take most pride and pleasure in designing the key experiments that will provide the decisive test of current theories. Such researchers rarely devise their own theories. What they do is carefully study and compare current theories, hence developing understanding of the key issues. They are then highly skilled in providing empirical tests of them. These researchers tend to be those with the opportunistic style, who will move on to new topics and issues at regular intervals.

Both research styles at their best are crucial to the development of our science. Those who originate theories typically also develop empirical research programmes to test and demonstrate their ideas. However, there is always a difficulty of partiality. Such theorists are often perceived by their colleagues to be somewhat biased in their choice of experiments, the interpretation of their data and, in particular, in their assessment of the merits of rival theories. Experimentalists with no particular theoretical preference, on the other hand, can provide enormous service to their field by the design of critical experiments and an impartial assessment of the ability of rival theories to account for them. In spite of this, I would have to say that in terms of fame and fortune, theorists and programmatic researchers generally fare rather better.

Grand theories

What I term 'grand' theories in psychology are arguably not really theories at all, but research programmes or paradigms. We could also call them meta-theories. However, we tend to use the term 'theory' when referring to them. Examples of grand theories in cognitive psychology would be the working memory model of Alan Baddeley and the mental models theory of Phil Johnson-Laird. Examples in social psychology would be cognitive dissonance theory (Festinger) and attribution theory (H. Kelley). All of these are 'theories' that have inspired very large research programmes; all have lasted for decades and survived numerous critiques and apparent empirical falsification. They are essentially paradigmatic in the Kuhnian sense, as their adherents share a belief system concerning the conduct and interpretation of studies.

One of the characteristics of grand theories is that they are more or less protected from refutation by their structure. In a sense, they are meta-theories that spawn specific theoretical accounts of particular phenomena. Hence, refutation and replacement of an empirically refuted child-theory does not do much damage to the parent. The rewards go to those whose names are associated with successful grand theories, such as the psychologists named above. By their nature, grand theories are hard to overthrow. However, they can be defeated through a process of Kuhnian revolution, as in the rejection of behaviourist learning theory in the 1960's, or sometimes just go out of fashion and fade away. These theories need leaders: respected figures who continue to inspire and motivate the research programme. They will typically publish books as well as journal articles. However, such leaders are rare individuals and not always replaced.

A grand theory with which I am well acquainted is the mental model theory of reasoning. This theory and its attendant research programme have a clear innovator and leader in Phil Johnson-Laird. It started with his book *Mental Models* (Johnson-Laird, 1983). In this book, Johnson-Laird challenged the orthodoxy (reflected, for example, in the work of Jean Piaget) that reasoning involves a mental logic comprised of inference rules, similar to those found in a logic text-book. Instead, he argued, people build mental models to represent possible states of the world. Valid deductions are those that are true in all possibilities or mental models. Hence, he argued, there is no need for logical rules. What you do is build mental models compatible with the premises of an argument and identify potential conclusions that are contained in these models. To attempt to prove their valid-ity, you can search for counterexample models that represent possible states of the world in which the conclusion does not hold. People are logically competent in principle, but often err in practice. For example, they may have insufficient working memory capacity to consider all relevant models, or their search for counterexamples may be biased by whether they believe the conclusion and so on.

Like most good ideas, it is essentially quite simple. Suppose you are given the following problem:

Jane is taller than Mary
Sally is shorter than Mary
Who is tallest?

According to mental model theory, you build models of the premises that represent states of the world consistent with them. In this case, you end up with a model that looks like:

Jane
Mary
Sally

where the top of the vertical array represents 'tallest'. Hence, the answer is Jane. In this case, there is only one mental model consistent with the premises. Consider the following premises instead:

Jane is taller than Mary
Jane is taller than Sally

Assuming no two persons of equal height, these are consistent with two different mental models:

Jane   Jane
Mary   Sally
Sally   Mary

Notice that you can still determine that Jane is tallest because this is true in both models. However, you could not answer the question, 'Who is shortest?'

Just considering this very simple application of the model theory of reason-ing, place yourself in the role of the experimentalist who likes to test theories.

How might we tell if people are indeed building mental models to answer such questions? Johnson-Laird (1983) wisely ruled out the idea, intuitive to some, that mental models take the form of visual images or any other consciously experienced and reportable form. They might give rise to images, but that is neither here nor there. So how can we test the theory? One idea, proposed by Johnson-Laird himself, is that problems requiring more models to solve are more difficult, due to constraints on working memory capacity. Another idea (Potts & Scholz, 1975) derives from the thought that different linguistic expressions result in the same mental models. For example, the first problem could be expressed as:

Jane is taller than Mary
Sally is shorter than Mary

Or

Mary is not as tall as Jane
Sally is not as tall as Mary

Suppose you present these premises and then a delay of several seconds before you ask the question: who is tallest or who is shortest? If they have been combined into a single (identical) mental model, then the linguistic form in which they were presented should have no effect on reasoning times, which is what Potts and Scholz found. (Note that demonstrating such null results can be problematic, see Chapter 9.)

We can see in this example the distinction between a grand theory and a specific theory or model that applies to a particular domain. The theoretical models require auxiliary assumptions to fit the characteristics of the particular task. Thus an early detailed implementation of the model theory was applied to categorical syllogistic reasoning (Johnson-Laird & Bara, 1984) involving a number of quite specific assumptions implemented as a working computer program and yielding, its authors claimed, some novel predictions not generated by rival accounts of the same task (Evans et al., 1993).

Of course, Johnson-Laird (1983) did not present mental models as a grand theory, but rather as a big idea. However, by the time of the publication of a second book (Johnson-Laird & Byrne, 1991) some eight years later, it already had the form and following of a grand theory. First, as this later book illustrates, the authors and their immediate collaborators had energetically applied the (grand) theory to a wide range of reasoning tasks supported by numerous experimental studies. Second, other reasoning researchers, especially in Europe, had become interested in the theory and had either provided their own experimental tests of it or else used model theory to provide *post hoc* accounts of particular findings. Finally, the theory had developed a clear enemy camp in the form of the (mostly US based) mental logic theorists, who still favoured the idea of inference rules as the basis for human reasoning. I do not think you could really call something a grand theory that did not have sceptical critics as well as enthusiastic supporters, and the model theory has plenty of both!

It is easy to think of theories that attracted huge followings in psychology (Skinner's behaviourism, Festinger's cognitive dissonance theory) but never without at least some sceptics and critics. I believe this is just as well. In other sciences, theories have attracted near universal support, with Newton's mechanics and Darwin's natural selection as good examples. Newton's physics, while vastly supported, proved to be wrong, or at least incomplete, when Einstein published his theory of relativity. With very high speeds or accelerations, Newton's principles will fail, but you could still land a person on the moon without needing the relativity equations. The problem was that with such a powerfully confirmed theory like Newton's, it took an exceptionally brilliant genius like Einstein to be able to see that it was not the whole truth.

In contemporary science, Darwin's theory is the nearest to having had universal acceptance that I have encountered. Natural selection contradicts and has hence discredited the theory of Lamarck that acquired characteristics are passed on to the next generation. It can apparently explain everything including the evolution of very complex structures like the human eye, which Darwin himself regarded as an extreme test case for this theory. I was invited to a meeting on evolution some years ago, attended by leading Darwinists such as Stephen Pinker and Leda Cosmides. It had something of the air of a revivalist meeting, with the presumably atheistic attendants sharing a common faith or certainty in the theory. And yet . . . could the theory be like Newton's, extremely persuasive but not the whole truth? It seems this may be so, as recent developments in research on genetics have apparently provided some evidence for Lamarckian transmission of acquired characteristics (Koonin & Wolf, 2009). Somebody always needs to think the unthinkable.

Grand theories are very powerful and can seem quite oppressive to researchers who do not wish to follow them. As a reasoning researcher, I have quite a lot of time for the mental model theory and have applied it in some of my own papers. However, in recent years, I have become convinced that one application of it, in explanation of conditional reasoning (Johnson-Laird & Byrne, 2002), is fundamentally wrong (Evans & Over, 2004; Evans, Over, & Handley, 2005). The difficulty this presents me with is that my argument (and that of my collaborators) is with the model theory of conditionals, not necessarily with the grand theory or meta-theory from which it derives. Yet a backlash of angry retorts from the committed followers of the mental models movement was inevitable (Schroyens & Schaeken, 2004).

Of course, the area we are entering here is the social psychology of science. Scientists conform to the same social psychological principles as everyone else. They form in-groups and out-groups, divide themselves into leaders and followers, apply stereotypes to perceived out-groups and so on. Particularly strong loyalties develop within tight-knit research groups but also within wider schools and movements such as the one inspired by the mental model theory. Social psychologists also discover independent individuals who follow neither one group norm nor the other. In my own case, I am neither 'in' nor 'out' where model theory is concerned. I see its merits in some applications and its weaknesses in others.

I have said above that originators of grand theories, schools or research programmes get the highest kudos of all. As an ambitious young researcher, you might demand of me in this book that I tell you how to devise one. Naturally, I have no real idea how one could set out to do this! The nearest to a grand theory that I have been involved in developing myself is dual process theory, which posits distinct processes underlying behaviour at an implicit and explicit level. This started for me as theory of reasoning (Evans, 1989; Evans, 2007a; Evans & Over, 1996; Sloman, 1996; Stanovich, 1999) but in recent years, I have developed something more like a grand theory that takes in the mind as a whole (Evans, 2008; 2010; see also Stanovich, 2011). The idea of dual processing has been developed and rediscovered in many fields and seems now to have attracted widespread interest. Neither I nor any one author can take all the credit for its development as a Big Idea but it is certainly out there as one now. The theory does have enemies as well, to whom Stanovich and I recently replied (Evans & Stanovich, 2013a; 2013b).

For my own part, the origins of this theory were anything but grand and grew gradually from some very specific and limited theorising about the findings in reasoning experiments that I was involved in during the 1970's. The progression from this to the two minds theory was long and arduous, with many intermediate publications along the way and much influence from the brilliant writings of Keith Stanovich (e.g. 2004) and other authors. The progression might not have happened at all but for the happenstance that I was invited to review two books on implicit learning (Berry & Dienes, 1993; Reber, 1993), a topic quite outside my normal area of reading. I discovered that these authors were working on dual process theories of learning, based on implicit and explicit cognitive systems. The links with reasoning theory (not mentioned in either book) leapt at me to the extent that I presented a rather surprised book review editor with a 5,000 word book review to draw out the implications for my own field (Evans, 1995). In a subsequent collaboration with the philosopher and reasoning researcher David Over, the general form of the theory was then developed quite quickly thereafter (Evans & Over, 1996), making sense, to me, of many years of research on the topic.

I guess one lesson this teaches us is of the importance of scholarship in research work (see Chapter 1). It is really quite hard to keep up with work in one's own research field without reading more broadly into possibly related areas of research. Yet it is the discovery of links and connections in one's broader reading that may be the essential catalyst, as happened in my case. I seriously wonder how different my subsequent research programme might have been had I not happened to review those books.

Testing theories

Ownership of a grand theory requires a combination of talent and luck that will evade most academic researchers. However, everyone who is successful in research will be involved with theory in order to motivate the design of particular studies and to facilitate the interpretation of their results. Here, it is useful to draw a distinction between theories and models, at least as I will use the terms. A psychological theory usually concerns the nature of mental processes in a particular domain. A model is an account of specific tasks and the behaviour associated with them. Hence, theories normally originate with thought about the psychological mechanisms, and models from attempts to provide specific accounts of empirical data.

For an early career researcher, it would be normal (and generally a good idea) to start by focussing on theories already found in the literature. Since you have no great emotional attachment to such theories, not having devised them, you might want to test them to destruction. The author of a theory (normally) seems happy with its empirical support, but you might find it implausible or insufficiently tested in some way. This is where you can happily adopt the Popperian role and ask of a theory what would prove it to be false. In general, this will be easier to do with models of specific tasks than with more broadly defined theories. Most psychological theories are typically not rigorously axiomatised so as to form a coherent logical system (see Chapter 8). There has, however, been an increasing fashion for mathematical modelling in recent years, as evidenced by many articles published by *Psychological Review*, the preeminent journal for psychological theory. If you have good mathematical skills, you should look carefully at following this trend.

The majority of psychological theories still tend to be implicitly statistical and, hence, hard to refute in the strict Popperian sense. As a result, there is often a gulf between theory and data. Theories are usually written in deterministic language, which is clearly an oversimplification, as experimental predictions and statistical analyses deal in behavioural *propensities*. The theory may talk as though frustration causes aggression, semantic similarity interferes with memory or authority figures are believable. But all of these will translate into propensities that are measured statistically when actual experiments are run. More frustrated participants will tend to behave more aggressively, but inferential statistics will be needed to test for significant differences between experimental and control groups. Word lists with semantically related words will result in distributions of errors with higher means than those with unrelated words, but the two distributions will normally overlap. On average, believability ratings will be higher when communications are uttered by high authority figures and so on. Almost never can we predict what a particular person will do on a particular occasion. Hence, logical falsification in the strictly Popperian sense is not available to experimental psychologists (see Chapter 8 – *the problem of probability*).

What then constitutes 'falsification' of a psychological theory for practical purposes? Conceptually, there are two distinct situations. One is that the theory predicts something should happen that is not observed in the experiment. The other is that the theory fails to predict something that is observed. In all, there are four possible situations that I label as follows:

Positive confirmation	Theory predicts X	X occurs
Positive disconfirmation	Theory predicts X	X does not occur
Negative confirmation	Theory predicts not-X	X does not occur
Negative disconfirmation	Theory predicts not-X	X occurs

As discussed in Chapter 8, scientists mostly use a positive testing strategy (Klayman & Ha, 1987). That is, they try to work out what a theory predicts will happen under given conditions and then perform a test to see whether it does. All the examples mentioned above test – in a statistical manner – positive predictions. (I will return to negative predictions below.) We induce frustration and predict increased aggression; we devise lists of similar words and predict increases in memory errors relative to a control; we use a high authority communicator and test for increased credibility of the communication. What we are actually doing is following what is known as the Delta rule of probabilistic causation (Lewis, 1981). Suppose we have two variables, A and B, such that A is a possible cause of B. The Delta rule expresses the relationship between conditional probabilities as:

$$\Delta B = P(B|A) - P(B|\neg A)$$

where ΔB is interpreted as a measure of the causal strength of the relationship between A and B. The idea is that in order for A to be a causal factor for B, the occurrence of A must *raise* the probability of B. The *extent* to which it raises the probability (ΔB) is thus interpreted as the strength of the causal relationship.

Actually, the Delta rule as stated is insufficient to define a causal relationship and may only capture a correlation. For example, I may notice in a gathering of academics that most of the older males are wearing ties. Specifically, I observe that:

$$P(50 \text{ years and older} | \text{wearing tie}) > P(50 \text{ years and older} | \text{not wearing tie})$$

but I should hardly conclude by the Delta rule that wearing a tie makes people older! Pearl (2000) deals with this kind of problem by introducing the *do* operator, thus distinguishing P(B|A) from P(B|do A) (for a less technical coverage of the basic concepts of causal networks, see Sloman, 2005). We can see that making someone wear a tie will not make them older. The do operator is actually implicit in the experimental method. In order to infer a causal relationship, we must first *intervene* by setting the variable A and only then infer causation by observing an increased probability of B given A. We must, of course, avoid introducing any confounding variables in the process (Chapter 3). We could restate the Delta rule as:

$$\Delta B = P(B|do\ A) - P(B|\neg A)$$

A scientific theory always involves causal relationships. In psychology, these usually involve hypothetical constructs and intervening variables. In Chapter 9, I will discuss some examples of biases in the assessment of probabilities that have been established in psychological research. A major tradition in this field is known as 'heuristics and biases' (Gilovich, Griffin, & Kahneman, 2002; Kahneman, Slovic, & Tversky, 1982; Pearl, 2000). The two terms in this phrase have a quite different status. The term 'bias' refers to a behavioural observation: a judgement that people make that deviates from some normative standard for correct reasoning. The term 'heuristic', on the other hand, refers to a hypothetical construct within the theory. Tversky and Kahneman argued that people judge probability using heuristics such as *anchoring*, *availability* or *representativeness*. These heuristics are causal factors in the theory but are not directly observable.

For example, the availability heuristic is the idea that we judge the frequency of events by the ease with which we can draw examples to mind. The fact that I can recall few examples of an event *causes* me to judge it to be an infrequent or improbable event. This is a level of causal reasoning different from that applied to the experiment itself, in which some independent variable A must be manipulated in order to demonstrate its causal impact on some dependent variable B. However, the experiment concerns the theory that provides the causal framework for interpreting the research. In the case of the availability heuristic, for example, Tversky and Kahneman (1973) provided a series of ingenious experiments testing positive predictions based on the availability heuristic. For example, suppose you read a list of names to someone and then ask them to judge how long the list was. Tversky and Kahneman predicted (successfully) that lists of names of famous people would be judged longer even though they actually contained *fewer* names than lists of unknown people. They argued that people would more easily encode the famous names and hence recall more of them. They also included a group who were asked to simply recall the names to show that this mechanism underlay the frequency judgement.

A theory like this actually sets up a chain of causal inference. In the above case:

> (do) include famous names in the word list → increased recall of names
> from list → increased judgement of length of list

where → represents a casual operator that we might read as 'increases the probability of'. Note that this chain does not capture the theory itself. The theory lies in the reason provided for the second causal link – the availability heuristic *explains* why increased recall leads to increased judgement of probability.

Most psychological theories use hypothetical constructs in this way to predict causal relationships. An example of an intervening variable arises with the frustration-aggression hypothesis mentioned above. Suppose we place two chimpanzees in a cage and leave some bananas behind a glass screen where they cannot reach them. We predict increased observations of aggressive behaviour between the chimps. We use 'frustration' as an intervening variable in the causal chain as follows:

> (do) out of reach bananas → frustration → aggressive behaviour between
> chimps

One way in which to disconfirm a published theory is to question the methodology of the experiments that purport to support it. For example, what is the right control condition for the chimp experiment? Suppose the original research compared this with conditions in which the bananas were in reach. The chimps would naturally eat the bananas. So we might argue that they were less hungry. This would lead to an alternative causal chain of inference:

(do) out of reach bananas → hunger → aggressive behaviour between chimps

Here, we have substituted 'hunger' for 'frustration' as the intervening variable in the causal chain. We have identified a confounding variable in the experiment that provides an alternative causal account of the finding. Note that we are not disputing the fact that placing bananas out of reach caused the increase in aggressive behaviour. What we are disputing is the causal interpretation of this finding provided by the theory. This would be the starting point for devising new experiments that would test the theory better. This does not have to result in actual falsification of the theory in order for the research to be publishable. It is sufficient to argue that the experiment was unsound and the original causal chain of inference unjustified. Our new experiments might show – with improved methodology – that the original claim was justified, but this is also a publishable finding. (Some rightly famous papers have, of course, introduced their big ideas despite the report of studies that were initially flawed.)

An interesting and important facet of this example is that the research will be publishable *whichever way the results turn out*. This is clearly the ideal situation, but will only be achieved by those with a strong grasp of both theory and methodology. It is not – by any means – the case with all experimental research that negative results can be published. For example, if you do an exploratory study to test your intuition that there is a relationship between A and B, it is unlikely to be published with negative findings. However, if we can demonstrate by strong argument that a published and influential theory *ought* to predict that A leads to B, then a finding that it does not will often convince journal editors that your paper contributes to knowledge (see Chapter 7).

There is an asymmetry between positive and negative findings. Events are by their nature *rare*; that is to say, for most A, $P(A) < P(\neg A)$. (Of course, there are exceptions such as the sun rising in the morning.) Thus, in the absence of a theory, to show that A causes B is of much more general interest than to shown that A does *not* cause B. When negation is used in natural language, it is generally to deny some presupposition (Wason, 1972). To say 'A is not the case' is anomalous unless used in some context where there is good reason to suppose A *is* the case. The statement denies A rather than asserts not-A. Thus you say, 'I have not been to the cinema for ages' to a friend who wants to discuss recent films, but not to one who wants to talk about sports. The same rules of conversation apply in the scientific context. This is quite apart from the technical difficulty of trying to prove null hypotheses true with a statistical method devised to prove them false (Chapter 9).

As illustrated earlier, positive testing can lead both to confirmation and disconfirmation. What I termed a 'positive disconfirmation' above is actually finding a negative result with a positive prediction. However, negative tests can, in principle, also lead to both confirmation and disconfirmation even though people find them hard to generate spontaneously (Chapter 8). A negative test is a prediction that if a theory is correct, then something should *not* happen. This actually involves another kind of causal reasoning that has to do with prevention. Suppose I know nothing about motor cars but notice, empirically, that the following rule typically holds true:

If you turn the ignition key, then the engine will start

However, I also notice that under certain conditions, this does not happen – for example, if the battery is flat or if the fuel tank is empty. These are *disabling conditions* that prevent A from causing B as it ordinarily would. Scientific theories can have prevention rules. In fact, certain scientific hypotheses are stated in such a way as to indicate this explicitly. For example, Hooke's Law states that:

The stretching of a solid body is proportional to the force applied to it, *within its limits of elasticity*

If you attach a spring to a hook and add weights to it, you will find that twice the weight induces twice the stretch – a neat example of linear physics. However, if you add too much weight, the spring will reach a limit and stretch no further – hence, the qualifying condition in the above statement. The law has a disabling condition. In general, we are looking at causal statements of the form:

(do) A → B, unless prevented by C

A theory that posits disabling conditions is open to negative testing. One apparent bias in statistical inference (Chapter 9) is that sample size may be ignored when making inferences from samples to populations. In their original paper on the 'representativeness' heuristic, Kahneman and Tversky (1972) made an astonishingly strong negative prediction. They argued that people would be completely insensitive to sample size when asked to judge the probability that a sample was drawn from a population, as size would not impact the perceived representativeness of samples. They presented some experiments in apparent support of this in which sample size had no effect on judgements of probability. A paper of my own was among the first to provide a *negative disconfirmation* of this hypothesis (Evans & Dusoir, 1977). We argued that they had used unnecessarily complex instructions that may have caused the negative findings. Using simplified instructions, our experiments – and those of later researchers – demonstrated that people did take account of sample size, contrary to the original prediction. Kahneman and Tversky (1982) later conceded that the evidence supported only the weaker position that people *underweight* sample size in judging probability.

Negative testing is fairly uncommon – I admit to having to search my memory hard for an old paper of mine where it actually occurred. Note that in the example cited, the rule about plausible negation applies. The prediction that people would ignore sample size was interesting because the normative theory of statistics indicates that it is highly relevant. In fact, cognitive biases can be defined either positively – as systematic attention to a normatively irrelevant feature of problems, or negatively – as systematic neglect of a normatively relevant feature (Evans, 1989).

I have talked in this section so far about the concept of testing other people's theories that are published in the literature. This involves understanding the causal structure of these theories and/or a critical examination of the methodology used to provide supportive findings in published studies. Setting up critical tests may disconfirm these theories, but it may also support them, and hence in the Bayesian manner increase confidence in them. Either way, the research contributes to knowledge and merits publication. Of course, the same applies to testing theories of one's own, which are likely to develop with more experience of research work. Some researchers are biased in testing their own theories, showing confirmation or belief biases, but this is by no means always the case. Others, however, are clearly seen to posit alternative accounts of their own findings and then try to find critical tests for distinguishing them.

Developing theory

I have indicated above that a priority for young researchers is to understand the causal structure of theories so as to derive both positive and occasionally negative ways of testing them. This will normally involve testing established theories that are published in the literature. However, this does not mean that you will not be *theorising* from an early stage. Publication in even middle ranked journals will generally require intelligent and convincing discussion sections in which the author(s) strive to achieve understanding of the empirical phenomena that they report (see Chapter 7). Hence, you will from the start need to engage in a quest to understand causal relationships, even if the output is less than would normally be awarded the accolade of a theory. However, it will be advantageous in due course if you can develop a well-formed theory of your own.

Actually, I think there are broadly two main approaches to how psychologists develop their theories. One is an extension of what I am alluding to above. Those with a strong empirical bent will be much engaged in the design of experiments or other empirical studies in which they analyse causal relationships. Thus, what starts as a set of hypotheses gradually builds 'bottom-up' into a theory. This happens by process of abstraction and generalisations. I am not sure exactly how Richard Gregory (1963; 1990) developed his 'misapplied constancy scaling' theory of visual illusions, but it might well have developed in the following manner. Consider the two well-known visual illusions shown in Figure 5.1. In the Muller-Lyer illusion, the line on the right looks bigger, and in the Ponzo illusion, the top line looks bigger, although in both cases, they are the same length. Do these two illusions have anything in common? Well, according to Gregory, they both contain features that the brain might interpret as perspective cues to distance. The left arrow could be the edge of an outside wall of a building, whereas the right arrow could be the inside edge of a room where two walls meet. These cues might

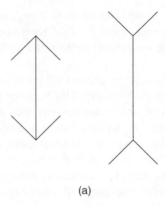

(a)

Figure 5.1 (a) Muller-Lyer illusion

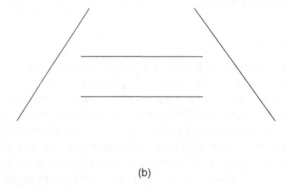

(b)

Figure 5.1 (b) Ponzo illusion

lead the brain to 'think' – by a default implicit processing mechanism – that the left line is closer to the viewer than the right. But if it is closer, then it must really be smaller to subtend the same visual angle. Hence, by misapplied constancy scaling, we get the illusion.

In the Ponzo illusion, the sloping lines could be perspective cues, like railway lines receding into the distance. In this case, the top line is 'further away' but subtends the same visual angle, and hence it looks longer. By inspecting these two illusions and thinking about their characteristics, we can abstract a common characteristic – perspective cues. We can then posit that the two illusions result from a common mechanism. The link is then made (by Gregory) with the well-known phenomenon of size constancy, in which more distant objects appear larger than their visual angle would require. The notion of *misapplied size constancy* is then formed as a simple theory with a hypothetical construct to explain causal links. Of course, the researcher is then required to examine a whole range of visual illusions for further supporting evidence and to conduct new experimentation. The theory is also open to falsification, as it posits a mechanism that goes beyond simple description of the phenomena.

While theories can develop in this bottom-up manner, many psychological theories develop top-down with thinking that is not at all attached to empirical studies. Such theories may develop, for example, from 'common-sense' observations of everyday life or from considerations of fundamental brain mechanisms, or by cross-disciplinary influences. One of the best examples I can think of is the field of evolutionary psychology, in which it is postulated that the architecture of the human mind and many of its specific functions are a consequence of the evolutionary processes that affected our ancient ancestors. I cannot see how such a theory could possibly develop bottom-up. No amount of thought about the results of psychological experiments could lead one to posit an evolutionary mechanism. The drive for this theory comes from evolutionary biology, and attempts to provide empirical test of its claims have met with scepticism from many critics. Unfortunately, as a commentator pointed out to me, psychological modules leave behind no fossils for psychologists to collect (Cosmides & Tooby, 1994; Cummins, 2004; Over, 2003; Stanovich, 2004)!

For most researchers, I believe that theory will develop as a natural consequence of good scholarship and good design of empirical studies. Experimental studies by their nature test hypotheses and investigate causal relationships. Striving to understand their results and to formulate the next well-designed study should always involve theoretical considerations, even when working in applied domains. In practice, this bottom-up thinking will link with top-down influences from your reading. You may be satisfied with the current state of theory in your field or you may find increasing reasons to doubt its adequacy as your research progresses. This may lead you to revise current theory or even to reject it and propose radical new alternatives.

KEY POINTS

- All researchers must develop *theory* in the sense of knowledge and understanding of the phenomena they study. Some will also wish to develop original theories and models
- Early career researchers normally start off by testing theories published in the literature by other authors. Some follow this pattern for their entire careers, focussing on designing studies to test and develop published theories
- Scientific theories build on causal relationships and typically incorporate hypothetical constructs and/or intervening variables. This causal structure renders them testable, in principle, by experimental means
- Theories may develop 'bottom-up' from attempts to explain observed phenomena or 'top-down' from prior thinking about causal

relationships. Researchers will normally develop theories later in their careers having first explored causal relationships through empirical research

- Grand theories, or meta-theories, are often referred to as 'theories' but are not testable in themselves. They are tested through application of more specific theories and models to empirical tasks. Such theories can be highly influential but are developed by few authors

6 Collaboration and supervision

It is possible to conduct research in isolation, but I do not recommend it, especially in a subject like psychology. Philosophers and theorists may work well on their own, but empirical researchers tend to work best in groups with much collaboration and interaction. However, there are some historical and cultural influences on research groups that should be acknowledged at the outset. It is sometimes said that teaching departments in universities follow either the zoo or the safari park model. In a zoo, you have one animal of each kind. I believe that 30 years or so ago, the zoo model dominated just about everywhere. Academic staff were chosen to teach a broad undergraduate curriculum, so the thinking was that you needed to have a pretty even distribution of expertise in all areas of the subject. In the UK, there has been a dramatic shift in recent years towards the safari park model, where you have a small number of packs of similar animals. Building groups of likeminded faculty has been seen as an effective means of chasing high ratings in the Research Assessment Exercise, which rewards strong research output with additional funding. Various solutions have then been found to the problem of delivering a broad undergraduate curriculum.

In the USA – with notable exceptions – the zoo model is probably still dominant, although there are still research groups. However, these tend to take the form of laboratories associated with individual faculty who have substantial research funding. Such groups may hence consist of clear leaders with a number of postdoctoral researchers and research students. Safari park groups also exist, however, especially where there are funded research centres in either industrial or academic contexts. (This is very common practice in continental Europe, for example, where elite researchers tend to enjoy full-time research positions, whereas their colleagues in teaching universities may have relatively limited support for research work.) These groups may be comprised of senior and experienced researchers, junior researchers who may be postdoctoral fellows or early career faculty, and research students, usually registered for PhD's. I worked for most of my career in such a group myself, which also extended to collaborators outside of my home university and to researchers who visited from time to time on sabbatical leave from their own institutions. The core group would meet every week or two to discuss research in progress, recent key publications in the field, rehearse conference presentations and so on. Everyone in the group, regardless of seniority and experience, found this enjoyable and helpful.

Research projects generally involve smaller groups, typically one to three investigators holding a research grant, with at least one research assistant or postdoctoral researcher employed to conduct the research. In our case, these subgroups associated with particular projects are usually formed from members of the larger group. You might collaborate with A on one project, with B and C on another, and with A and C on a third and so on. Working at its best, the larger group is following a research programme within which the particular projects have a role. Sometimes, such larger groups are identified as research centres that may be programmatically funded by an external body, or that may be funded by a set of individual project grants. Some universities in some subject areas have formed still larger groups known as research institutes. An institute would be comprised of a set of research groups or centres with some high level programmatic connectedness. These may be formed to support interdisciplinary endeavours such as cognitive science, neuroscience or evolutionary studies.

My advice to PhD students and early career researchers is that you should try to locate yourself in larger research groups and centres if this is an option open to you. Although I have been fortunate enough to work with a reasonably large group in recent years, this was not always so. Early in my career, collaborative possibilities were much more limited, although I usually had PhD students to work with. Both from my own experience and from observations of others, I can say that both research students and early career researchers do seem to benefit enormously from involvement in programmatic research groups, while those working in comparative isolation – no less able – can struggle to make progress. Of course, you want to ally yourself with a group that is not merely large but demonstrably successful in attracting funding and in publishing in high quality outlets.

What precisely are the benefits to young researchers? The first area concerns the acquisition of knowledge and expertise of your chosen field. Although scholarship (see Chapter 1) in the sense of studying the literature is essential, it is hard to fully grasp a field in this manner. I have occasionally met researchers from foreign countries who have enthusiastically followed and read the main literature in my own field for some years. However, they have not had their PhD supervised by an expert on the topic, nor have they had the opportunity to collaborate with any established researcher in the field. They have lived somewhere remote from the main centres of gravity of the field with few opportunities to attend the meetings in which major participants present their work. The result seems to be that, although they are acquainted with the main concepts, they rarely understand properly how to do research on them. I guess this brings me back to the purpose of the current book: you do not learn how to do research by reading the literature. The best way is to work with people who already know how to do it.

As I mentioned in Chapter 1, senior researchers – who have worked in the same field for many years – acquire an understanding that is difficult for junior researchers to match. They have not just read a lot of literature – they have lived through it. It is one thing to 'know' a field by reading the research reported in the past 20 years. It is quite another to have read the papers as they came out, followed the twists and turns in theory and method and to have been an active participant in the unfolding story. Early career researchers who are supervised by such experts, or who work with them as junior collaborators, will gain immensely from this know-how and perspective. There is no short-cut to mastery of a field of scholarship but development will certainly be quicker in the company of established scholars. However, as I also mentioned, elder researchers can develop fixed views, so it is important that the younger members contribute some independent critical thinking. As one of the mature guys, I can attest that the fresh thinking of junior members of a research group is great benefit as well.

Working with groups also conveys a lot of practical advantages. Big groups tend to attract regular research funding so that a PhD studentship may lead to one or two postdoctoral contracts, enabling the young researcher to establish a base of publications from which to launch their academic career. Such groups also tend to be networked both nationally and internationally to other similar groups in other universities. The senior researchers will receive manuscripts reporting significant new work well before they are published. Experts from other universities will visit regularly, giving talks, and sometimes extending their periods of visit with sabbatical leave. Members of the group, or their closely networked external contacts, will organize specialist conferences and scientific meetings in which you will be involved. The university library will carry the right stock of books and journal subscriptions (now mostly electronic) to support your field of research. Specialist equipment and laboratories, if required, will have been established. And so on, and so forth.

Although an enthusiastic supporter of groups, I should acknowledge that there are arguments against them. One view is that groups encourage conformity, 'group-think' and confirmation biases. I guess it depends somewhat upon the overall culture and personalities involved. I certainly know of cases where group leaders appear to be very powerful and controlling individuals. This is particularly likely to occur where only the group leader has a tenured or securely employed position. In the groups of likeminded faculty that I am familiar with, there are several permanent members (excluding the short-term researchers and students) who hold secure positions and only work together because they choose to do so. These genuine, relaxed collaborations can be very productive.

It may be that the opportunity to work in this kind of environment does not come your way, at least not early in your career. What is to be done? It is clearly important to attend conferences and present papers on your own work. In my experience, smaller and more specialised conferences or scientific meetings are much more useful than large-scale gatherings of grand societies. You must make efforts to make contact with researchers in your field who are located in other institutions. You can email them with requests for comments and guidance; sending draft manuscripts and so on. The best way to get known if you can – despite your impoverished environment – is to publish some papers that make a significant contribution to the topic. You will quickly be noticed, especially since established experts will be sent your papers for review by the journals.

It is also possible to establish collaboration at a distance, something made much easier by modern communications. For example, two of my major collaborators in the later part of my career have been David Over, a philosopher/psychologist working at Durham University some 400 miles from my home institution in Plymouth, and Valerie Thompson, who works in Saskatoon, Canada, thousands of miles away. All the time that I have worked with David, we have lived in these two cities and we have never had a period of extended study leave where we were in the same place. In spite of this, we have managed to write two books together, as well as a number of journal articles, and to receive collaborative research funding from external agencies on several projects. How is this possible? In our case, we discovered common interest through our publications, made contact initially by telephone and email, then met and talked at conferences and so on. Once established, the collaboration has been remarkably easy to maintain. In Valerie's case, our collaborations have been facilitated by her regular sabbatical visits to Plymouth, as well as regular meetings at international conferences. But we remain in constant touch by email.

There is nothing at all unusual in such remote collaborations. One of the most famous and enduring collaborations in psychology is that of Danny Kahneman and Amos Tversky. They were both native Israelis, which is where they first met and collaborated. However, each spent the bulk of their careers working in the United States but never in the same university. In fact, I can think of quite a number of similar examples in my own field of people who have sustained long-term and productive collaborative research programmes without ever living or working in the same place. Even if you live on different continents, it is easy and free to talk via web based communication systems. Of course, email is an almost ideal medium of communication for research collaboration. In fact, you tend to use it a lot with collaborators in your own department who might just as well be on another continent so far as this means of communication is concerned. Of particular benefit is the ability to exchange attachments of word-processed files. Providing you use compatible software – which most people do these days – you can send drafts of joint papers up and down, editing and re-editing each other's writing with great ease. (There are now also means to share files through the cloud directly with colleagues if you can be bothered to use them.) Although it is not quite the same as having a collaborator in your own institution, many of the same benefits accrue. And of course, for logistical reasons, it may simply not be possible for you and your potentially most fruitful collaborator to work at the same address.

Email communication should by no means be restricted to those with whom you have a formal collaboration (one based on joint funding or likely to lead to joint publication). It is a more effective way of developing and exploiting networks of contacts with related interests than conferences because actual meetings occur relatively infrequently. There are senior researchers in my field whom I have known for many years but have rarely or never engaged with in a formal collaboration. Yet, in some cases, we exchange emails on a quite frequent basis, discussing issues of common interest, commenting on each other's papers, and even challenging the other to develop some new aspect of their theory or to run some new experimentation to test their latest ideas. Sometimes, I think of an experiment that I believe will provide a critical test of someone else's theory.

I may then email them with the design and ask for their comment on my prediction before I collect the data. It helps if you have met the individual personally, but it is not essential. You can have effective email relationships with people you have not met in the flesh.

Rules of collaboration

Lest I have made it sound idyllic, I should admit that collaborative relationships do not always work well and can sometimes lead to significant problems. Clearly, personalities and egos can clash and collaborations that look ideal on paper do not always work in reality. Although there may be no explicit rules of collaboration, there are some implicit ones of which you should be aware.

Collaborations are rarely entirely equal. Usually, there is a lead researcher for any project and one or two others collaborating on this. The leader is not necessarily the more senior researcher and not necessarily the same individual on different projects. For example, I have collaborated with the same people sometimes on one project where I take the lead and sometimes on another where they have this role. The leader is usually the one who has suggested the project in the first place and would be the named principal investigator on a grant application leading to the funding. The leader will tend to take the initiative in organizing research meetings, drafting the papers and so on. However, it can be more complicated than that.

Professional research has to lead to publication, normally in the form of refereed journal articles. Although other disciplines have different conventions, in psychology, the first author on a paper is regarded as the senior author and accrues some tangible benefits from this. First-authored papers make more impact on your CV when you apply for jobs or promotions and give you more status with the reading public for your field. They are also generally easier to assess in calculations of citations to your work in the literature. Hence, first authorship is a clear benefit for which an appropriate cost must be incurred, if collaborative relationships are to survive and prosper. In fact, the major rule of collaboration that I would advocate is that order of names on publications – especially the first authorship – should be established as early as possible in the research.

It generally works best if first authorships alternate (or rotate) among regular collaborators. Hence, it can be agreed within a research project that there will be two or more papers to come out of it and to know at an early stage who will be leading on which papers. Strictly speaking, the rules of authorship are not to do with the amount of work that individuals do on the research, but rather to do with their intellectual contribution. For example, leading on the development of hypotheses and design of the empirical studies provides a major claim to first authorship. This will usually lead to the expectation that the first author will do the main work of drafting the paper for publication, with co-authors contributing comments and revisions. A further issue concerns the authorship rights of research assistants who may be employed to carry out the research work. If their role is confined to collecting and analysing data, and they make no effective contribution to either design, interpretation or write-up, then they are strictly speaking entitled to no more than a footnote, however hard they may have worked. Most research-ers, however, tend to err on the side of generosity when deciding which names to put on a paper and should be sensitive to the career needs and ambitions of their research assistants.

In practice, collaborative research teams are often unequal in the assignment of first authorships: you see a number of papers by Bloggs and Jones and few where Jones is the first author. This may happen because one of these research-ers is regularly taking the lead role in the research. You may find, however, that the second author in such cases is the lead researcher in other lines of work not involving the first. So, Bloggs and Jones publish regularly on one topic, while Jones and White publish on another. If people are happy to work this way, then this is fine. However, sometimes the dominant first author is just a much more fluent writer and by tacit agreement ends up writing most of the papers. Slow writ-ers often prefer someone else to take it on, and fast writers hate waiting for slow ones to produce their drafts! However, if the output does not fairly represent the balance of the research team, then the fluent writer may have to consider helping their colleague to achieve more first authorships by writing more of the draft than would be usual for a co-author.

Authorship disputes are the most likely problem to arise in collaborative rela-tionships, and departments may consider issuing formal rules of guidance on this. For example, in my own department, some years ago, we issued guidance on the publication of PhD work. We stated that such work should normally be published in joint names of student and supervisor and that the student would normally be the first author. This guidance was intended to protect both parties. If supervisors are doing their job properly (see Chapter 6), then their contribution should be worth an authorship. (A practical point also is that departments like to see some tangible benefit for their staff from an investment in research students who move on after a few years.) At the same time, research students need to develop their careers, for which first-authored papers are particularly important. However, it is not difficult to see the problems that can arise with this rule. Supervisors are, in the nature of things, more experienced scholars and writers and may be able to write up a paper more effectively and get it published more quickly or in a better journal. You may need a supplementary rule that theses can be written up by the supervisor as first author if the student is unable or unwilling to write the papers within a reasonable time period.

Trickier to handle than authorship disputes, in my experience, are disputes about intellectual property and ownership of ideas. In the ideal research group environment that I have described above, there is much interaction and exchange of ideas going on all the time between members of the group. This occurs not just in organized seminars and formal supervision meetings, but in corridors and over coffee. The downside of this is that it may be difficult for individuals to know where particular ideas originated. Ideally, no-one should care, but the people who feel particularly sensitive and vulnerable are the students and junior members of the group. Sooner or later, one of the senior researchers will write something in one of their papers that the student will believe was their own idea. I think this is almost inevitable. As a research student, I felt that my own ideas were once used in this way by a senior colleague. As a senior researcher, I have occasionally been accused by a junior colleague of doing the same to them.

If it should happen to you in a junior role that you feel your ideas have been appropriated by a senior colleague, do not assume that a breach of integrity has necessarily occurred. The first recourse is an amicable chat with the supervisor or colleague concerned, not an angry letter of complaint to the Chair (or Head) of your department. Most academics are honest and fair and do their best assiduously to cite the sources of their ideas and give credit where it is due. However, it is a psychological impossibility to know precisely where all one's ideas come from. Generally speaking, junior researchers derive great benefit from the knowledge and skill of their senior colleagues, as indicated at the start of this chapter. However, senior researchers also benefit from the fresh thinking and new ideas of their students. They are not likely consciously to use these without attribution but it may happen accidentally.[1] The best protection for the student is to get into the habit of writing their ideas down as early as possible. Then you can take your dated report along and show that you had first developed the idea; this will simplify matters greatly. There is also another possibility to consider. The supervisor may have given you the idea in the first place without you realising it. They may even have done so intentionally. This leads me on to the art of PhD supervision.

Supervision of research students

Once a young researcher succeeds in finding a teaching position in a university, they are likely quite early on to find themselves supervising PhD students, even if their own PhD was quite recently completed. Working with PhD students can be stimulating and rewarding and helpful to the progression of your own research programme. This should certainly be the case if the student is bright and working on a topic close to your own expertise and interests. However, be warned. Working with weak students can be a totally draining and demoralising experience and supervising students working outside of your own research field will be difficult and very hard work.

As I mentioned in the Introduction, PhD supervision practices in the UK have improved greatly over the past 30 years.[2] This is mostly due to pressure from funding bodies for reasonable completion times for the students they support but has also been assisted by the issuing of guidelines for good supervisory practice from professional associations. These days, many departments also assign independent individuals or small committees to monitor the progress of research students and to pick up any difficulties in student-supervisor relationships. A consequence of this (certainly in the UK) is that the concept of what a PhD is has changed quite a lot. Students used to think of a PhD as a grand work, making a major contribution to their discipline, no matter how many years it took to produce it. Nowadays, the emphasis is much more on research training, almost as though the student was an apprentice to the research supervisor. Nevertheless, most university regulations still require that a PhD makes a significant contribution to knowledge in its discipline; ensuring that it does so places a great burden of responsibility on the supervisor.

Where one-to-one supervision is practised, the relationship between supervisor and student can become close and intense. It is also a situation in which both parties are vulnerable if things go wrong. Early in my career, I once had a student who was very personable and articulate and always to be seen carrying the latest impressive book under his arm. Everyone who knew him superficially assumed that he was an excellent student. As his supervisor, it dawned on me after a few months that he was a quite hopeless researcher. He was one of those students who can only ever deal in generalities and could never get to the level of specificity required to design an empirical study. Eventually, I refused to support his progression to a new year of funding, after which the sky fell in on me. Colleagues were sceptical of my decision, especially given my inexperience in supervision. Worse, the student had a very influential father (a senior civil servant, if I remember correctly) who wrote to the boss of my institution with a formal complaint. I was only vindicated when the student was reassigned to new supervisors who discovered exactly the same weaknesses that I had been grappling with.

Of course, students are very vulnerable in a one-to-one relationship also. Supervisors can be poor: selfish, inattentive and neglectful, for example. A student who suffers may feel that they have little recourse, especially if the supervisor enjoys high status in the department. As a result, many universities now practise joint supervision with at least two supervisors and also have some independent individual or committee responsible for monitoring progress. In spite of this, it is quite common for the student to end up working very closely with one of the supervisors, particularly at the writing-up stage. Students who have such problems with a supervisor should try to get themselves re-assigned well before this stage. After all, that is one of the reasons that a monitoring committee should have been set up.

The ideal situation for a research student with ambitions of an academic career is to complete their PhD by the end of their funding period and to have two good papers in press with learned journals at the same point. Then they are well placed to get themselves a postdoctoral fellowship, which will enable them to build further publications towards the kind of CV that will lead to a junior appointment in a strong university department. In my experience, few students achieve this ideal. Most take longer than the time originally allocated and distressingly few have papers in press at this stage, although most will have given conference papers. I say 'distressingly' because the time allowed for the PhD work is quite sufficient to achieve these goals, provided that it is successfully managed. Helping the student to manage the time is a key role for the supervisor.

In the following paragraphs, I describe my recommended approach to the supervision of a PhD. In the UK, students who are fortunate enough to have full-time funding may complete a PhD in three years of study. In the USA and other countries, PhD's may take more like four to five years to complete, as students need to fund themselves by part-time work as teachers or research assistants. This pattern is becoming more common in the UK also, as the number of fully funded places is reducing for a variety of reasons. Wherever you are based, a PhD will involve a large and substantial piece of research (normally empirical). Research students will probably not have had any previous experience of a project of this size and scale and it may seem very daunting at the outset. In my view, one of the most important tasks for the supervisor is to impress on students in the early stages the *urgency* required to complete their task. At this stage, students commonly feel that there is plenty of time and no reason for any hurry. They must be wrong in this belief (!), as all the evidence – as I have already pointed out – is that most students fail to complete the task on time.

The main way in which you can help your student is by structuring the task and decomposing it into a series of sub-stages with a specified time scale. A task such as 'complete a PhD in n years' is impossible to get a handle on. Tasks such as 'complete a literature review of field X in three months' or 'run and analyse two experiments by the end of your first year' are much more amenable. Nevertheless, we all suffer from the 'planning fallacy' (Bueler, Griffin, & Ross, 2002), a chronic cognitive bias to underestimate in advance the length of time that any task will actually take. For this reason, margins of extra time need to be allowed in any planning process.

I sometimes find it helpful to work backwards from the end goal in order to convince a first year student of the urgency. In the example that follows, I assume that a three year period of full-time research is available and about eight experiments, or the equivalent, is the expected scale of the empirical study. If you work in a system where different parameters will apply, then please recalibrate as required. The principles are the same. Hence, we imagine that in three years' time, the student will have finished the thesis and written two journal articles submitted for publication. What is the date by which all collection of data should be complete? In my experience, you should aim for this to be at least six months before the end of funding, even with a continuous writing approach (see below). So, as Newell and Simon (1972) propose in their famous work on problem solving, we have created a sub-goal by working backwards: complete experimental work in 30 months. Now we take the sub-goal and decompose that in the same way. How long will it take to design, run, analyse and write up a technical report on each experiment? Say three months. We already have an outline shape of the three years:

Months 1–6	Unplanned
Months 6–12	Two experiments
Months 13–24	Four experiments
Months 25–30	Two experiments
Months 31–36	Final write-up

At this point, the student may frown and start to appreciate that there is not quite so much time available as she (or he) first thought. Now what has to be done in months 1–6, she asks? Quite a lot, actually. In this period, she must decide the hypotheses to be tested and the methodology to be employed. To this end, she must conduct systematic reading and reviewing of literature relevant to the project and conduct pilot experimentation. I would keep the process of working backwards and decomposing the tasks to the point where the immediate tasks (say for the next two months) are clearly specified and planned. The advantage of doing it this way is that the student (a) grasps the urgency of making progress right from the start of their funding period and (b) sees the purpose of the initial tasks in the context of the overall three year project. You may come down to specifying that the first two months is spent in reading around the chosen topic area, specifying hypotheses and drafting an outline review of the papers read. Of course, you could have simply specified this as the first task without going through the task decomposition process that I have suggested. However, the student is then simply travelling hopefully, with no real idea of how she will arrive and no context with which to enforce the urgency of the work.

In my view, the biggest mistake that students and supervisors can make is to allow writing up of the work to be deferred. It is essential that students write up reports on everything they do as they go along. If they conduct an experiment, even a pilot study, then they should produce a technical report with fully specified method and results sections (brief aims and conclusions at the start and end will suffice). This should be done immediately on completion of the experiment. If they read a set of papers, then they should produce some kind of written review, even if rough or brief, again on immediate completion of the task. There are several good reasons for this. First, the student gets into the habit of writing and practises this skill (with the feedback you will provide) from the start. Second, the process is efficient. Technical reports will include details of methods and comments on results that could well be hard to recall if the work is written up from notes months (in some cases, even years) later. Literature reviews written at the time will need to be edited and expanded later, but much re-reading of papers will be avoided. Third, the student will not be faced with the huge psychological problem of facing a mountain of writing in their final year with no base to work from.

A weakness in the UK system compared with that used in North America and some other countries is that students are often encouraged to write long theses that then have to be scaled down for publication. It is quite common, when examining theses, to find experimental studies reported at a length that would never be permitted in a journal. Students often report excessive amounts of data analysis that are either redundant or that focus on aspects of the data that are uninteresting and will receive little discussion. However, while the system permits this approach, it does not *require* it. One consequence of the traditional approach is that the reports will have to be re-written in a quite different way in order to achieve publication. Thus, the student has (a) learned a method of writing up research that will need to be unlearned, as it will never be of use to them again and (b) has made the production of journal articles during the funded period of PhD research very difficult.

In actual fact, there is no reason why you should not be teaching your students to write journal articles from the start (see Chapter 7). It is much more efficient and effective for the student to write up their work as they go as articles submitted for publication and then to turn these articles into experimental chapters of the thesis. If this results in shorter, sharper and better focussed experimental chapters, I can assure you that no dissertation examiner is going to complain about that.

In my experience, achieving the optimum output within the funding period is hard work for the supervisor as well as the student. All PhD students are by definition novices at what they are doing. Many have false beliefs about the task, often reinforced by the culture of research students in the department. All of them (as well as you) suffer from the planning fallacy. They may be distracted by personal problems (break-up of relationships is a very common fate for research students). I still think that you can help them a great deal by following the principles I have suggested here, which in summary are (a) to structure and decompose the tasks into manageable units with identified time scales, (b) to require a continuous write-up of all work immediately following its completion and (c) to teach them to turn journal articles into theses rather than the other way around.

KEY POINTS

- Collaborating with others is of great benefit to researchers, especially those engaged in empirical studies. It is much easier to design and interpret studies when talking them through with colleagues of similar interests and skills to yourself
- Working in a research group with common interests in one location is ideal. Bigger and better known groups will have many advantages, such as regular research funding, many academic visitors and the ability to pool knowledge and contacts for the benefit of all
- Where this is not possible, or even when it is, remote collaborations with colleagues in other universities is recommended and quite feasible. There are many examples where this is successful even when quite large distances separate the collaborators. Regular contact by email together with occasional meetings at conferences and so on can sustain an effective working relationship
- Collaborations can also lead to problems and it is good to establish early on who will take the lead in writing papers and assume the advantages of first authorship. Where PhD students are involved, departmental guidelines may be necessary. Sometimes, disputes arise over authorship or, more generally, issues of intellectual property. In particular, junior members of a group may feel their ideas have been used by senior members. I discuss precautions that can be taken

- PhD supervision can be very rewarding but is also a very skilled and demanding task. All PhD students are novices and tend to overestimate the time available. It is essential for the supervisor to ensure a sense of urgency and also to structure the task into much smaller and more manageable units. I strongly recommend that students be required to write up all their work as they go along

7   Communication of research

A useful definition of academic research is that it is work that advances knowledge *in the discipline*. It is not to be confused with the everyday usage of gathering knowledge for yourself, as when a journalist researches a piece she is writing for a newspaper, or when a prospective house-buyer researches the market, or when a producer of a TV documentary researches a scientific topic. In the everyday cases, the knowledge is normally already available (to someone) and the individual is gathering it for their own use. In academic research, the knowledge gained must be original to the subject. One consequence, discussed in Chapter 1, is that your scholarship must be strong enough to demonstrate the originality of the work in the context of the published literature. However, this definition also implies that your work, in turn, must be published, so that it adds to the body of knowledge available to all scholars of your subject.

I sometimes hear research students or junior colleagues talk of their work as though they had completed the research but not yet written it up. To me, this is a contradiction in terms. The research is not complete until it has been written up and, in fact, published. Otherwise, it is adding to no -one's knowledge except your own. Also, until you write the work, you have not yet attempted properly to structure it, place it in the context of the research literature and provide a theoretical interpretation of it. Nor has it undergone the rigours of test by peer review. It never ceases to amaze me that some people think that this process is not part of the research. You might just as well collect data and not bother to analyse it, as far as I am concerned. The truth is, though, that a lot of researchers find writing difficult, the least enjoyable part of research, and are apt to put it off as long as possible. My response is to say that if you are going to perfect just one of the skills involved in running research, it had better be writing because without published output – preferably in journals of high standing – you are essentially invisible to the academic world. I will therefore concentrate for the most part in this chapter on the skill of writing and the approach to securing publication in refereed journals and other published outlets. However, research is also communicated – to a limited extent – by a range of oral presentations, which I discuss first.

Oral communication of research

Oral presentations – conference papers and research seminars – have an important place in academic output. However, you should never fall into the trap (as many do) of thinking that conference presentations are effective means of communicating your research to your discipline. First of all, the number of people who attend a conference *and* attend your paper *and* take in what you are saying is a tiny fraction of the number you can reach with a published article, even in a journal of modest repute. Second, they will absorb little of the detail without a written text to study (a conference abstract, even a long one, will hardly suffice). By all means, attend and contribute to conferences, but do not delude yourself into thinking that the research is now communicated and finished if it has not also been written up for publication.

One kind of conference is the large gathering of psychologists from all over the discipline. The conferences organized by the American Psychological Association, the British Psychological Society, the European and World Congresses of Psychology and so on come into this category. Quite why such conferences survive and prosper is a mystery to me. I occasionally take part in one of these, but only if I am invited to participate in a symposium focussed on a research topic with which I am involved. I personally dislike these meetings, which I find big and impersonal and much too broad in scope. There are few papers that I can attend that will have any impact on my own research work, even if they are well presented. Also, the pressure of timetables at these events has constrained papers to be ever shorter. The norm nowadays seems to be a 20 minute slot, which means really 15 minutes by the time you allow for setting up your PowerPoint presentation and taking one or two questions.

It is likely that early in your career, you will be confined to one of these short slots at a major professional meeting. You really cannot prepare a short enough talk for this time period. Almost everyone will prepare too much and end up rushing through detail that will baffle rather than enlighten the audience. You just have to believe how little you can get across in such a time period. The rate of overheads per minute varies greatly between speakers, as some people like to write all their points and others to use brief mnemonics. However, you can extrapolate from your experience of teaching undergraduates. If you find – say – that you need an average of 24 overheads for a one hour undergraduate lecture, then you should allow yourself just six for the 20 (really 15) minute conference slot. It is no good writing 12 and hoping to fit them in. You can, of course, rehearse in private and time yourself. However, be warned that when you give the real thing, it will take longer. In a real social situation, you expand points to respond to the non-verbal signals of your audience.

The only thing I like about the 20 minute slot is that it is less boring when I am the listener! I find the majority of conference papers tedious and unintelligible, so at least I know they will soon be finished. The presentations that work will try to make no more than two or three points and describe no more than one or two studies. However, they will have the same story structure that I will discuss later for refereed journal articles. That is, they will have a beginning – framing of the research question, a middle – description of the empirical study, and an end – interpretation of the findings. Humans have a love for the narrative form that is arguably innate to the design of our cognitive systems (Stanovich, 2004). All successful communications of scientific research tell a story.

What function do big conferences serve for the young researcher? Basically, they help to get you known. To this end, it will help if your presentation is comprehensible and reasonably interesting. However, the main benefit is probably to be derived from the informal side of the event. Get a senior colleague from your own department to introduce you to people from other universities, especially influential ones like senior researchers and journal editors. If you have published some papers, you may be rewarded by seeing one of the established authors visibly putting a face to your name, indicating that they may have read your papers or at least noticed them when scanning some journal. And remember the golden rule I gave you in the Introduction about big name, big ego researchers – always talk to them about their own research. It helps if you have actually read their work, but they will probably be so busy expounding its finer detail to you that they will not notice.

As a result of talking at such meetings and publishing two or three articles in reasonably visible journals, you will likely start to get invitations to talk at departmental research seminars or colloquia. Most psychology departments organize regular programmes of talks for external visitors and put aside a modest budget for paying travel expenses. It is quite common for early career researchers who impressed somebody in the department at a recent conference to receive such an invitation. Here, the timescale is much more relaxed, as you will get the 45–60 minutes normally reserved for major keynote speakers at the big conferences. The atmosphere should hopefully be relaxed also, although one or two departments may have a notorious individual who likes to ask aggressive questions.

With a longer time-slot, you can present the equivalent of one or two published papers with enough time to provide useful detail. However, do remember to tell the beginning and end parts of the story. Some young researchers (and even sometimes experienced ones) will give a talk that is all middle. That is, they dive into experimentation with little preparation and spend most of the talk describing the fine detail of their statistical analyses, stopping only when the last higher order interaction has been sufficiently explained. Of course, to some extent, this will reflect the research styles discussed earlier in this book. Empiricists are wont to describe as many of their elegantly designed experiments as possible in the time available, while Theorists will concentrate – sometimes exclusively – on the bigger picture.

Unless you are invited somewhere with a very large research group in your own field, departmental seminars will tend to be presented to a fairly general audience. Whether writing or speaking, empathy with your audience is an essential skill for effective communication. Whatever the level of the audience, you must prepare the ground. I will talk later about the importance of opening paragraphs in journal

articles. The opening five minutes of a seminar presentation is the equivalent –
if you lose the audience here, you are most unlikely to regain them later in the
talk. Contextualisation is the key. Or to put it another way, you need to motivate
your audience (or readership). In all probability, the middle part of your talk (the
equivalent of method and results in a journal article) will be quite technical and
hard work to follow. Whether or not your audience will make the required effort
will depend in large part upon whether you have motivated them to do so. So
provide a broad context at the beginning that people can understand, even if they
are non-specialists.

Some topics are intrinsically easier to sell than others. If your speciality is in
social cognition and you are reporting a study of stereotypes, then you have a head
start. Everyone in the audience is a social being with experience and possession
of stereotypical behaviours and attitudes. Hence, you should be able to engage
their interest early on with some real-life examples they can relate to, even if their
own specialism is in neurobiology or psychophysics. If, on the other hand, your
own topic is psychophysics, you have a rather harder job to engage the general
(psychological) audience, but that does not mean you should not try. In my experience, some speakers can make a talk on social psychology sound dull and others
can bring to life the most technical of topics. In all cases, it will help to follow the
structure of the well-formed journal article (Figure 7.1) that I will discuss later
in the chapter.

The other opportunity you may get to present your research orally is at a specialist meeting or workshop. The ideal meeting (in my view) is one where specialists in a research field get together for a couple of days and – barring the presence
of some students and non-speaking collaborators – everyone is both speaker and
audience in turn. There are no parallel sessions, so you all share the same experience. Such meetings are usually by invitation only, so you need to make your
mark in publication first. Normally, you get the longer format (40–60 minutes plus
discussion). Although the same general rules apply, it is much easier to present
your research in this context since the audience will generally be familiar with
the basic literature, the issues and the methodology. The one potential pitfall is
that your talk as prepared may have close connections with some of those that
have gone before, so you need to be flexible. You should be prepared to skip slides
where the method or theory presented was covered by a previous speaker and
draw theoretical connections to other papers where they become apparent in the
course of the meeting. It is rather bizarre – as sometimes happens – to hear someone stick rigidly to a prepared script, ignoring the context around them. It is also
unsympathetic to the function of such meetings, which is to develop a collective
consciousness of key issues across the whole period of the workshop.

The process of writing

As I said at the outset, the key communication of research is through its various
forms of written publication. Before discussing particular forms, of which the most
demanding and difficult is the empirical journal article, I will make some general

comments about the process of writing. The first point to note is that individuals seem to differ greatly in how they write. My old supervisor, Peter Wason, who was known as a fine writer of clear English, actually produced his work by a laborious process involving numerous drafts and re-drafts. He used to describe this as a 'dialectical' process that improved his thought and understanding of the issues. However, it was not very efficient and Wason was not all that prolific by modern standards, producing about 40 publications in a 25 year period, just two of which were authored books.

I do not quite know how I learned to write or when it became as comfortable a process as it is nowadays. I do remember that my writing style as a PhD student was quite poor at first and that I had numerous criticisms from Wason (whose first degree was in English Literature) to contend with. Apart from this input, my own writing skill is entirely self-trained, as I guess is most people's. Nor do I know how to teach someone to write the way I do. In complete contrast with Peter Wason, I never write outlines or rough drafts. Instead, I carry an idea for a paper around with me for weeks, doing some conscious or unconscious work on it but not writing down so much as a single note. When I feel ready to write, I sit at the word-processor and it flows out in near final draft form more or less as fast as I can type it. The process is tiring though, so I rarely write for more than 3–4 hours on a given day (finding something else to do with the remaining hours!).

I have never met anyone else who appears to write as I do, but discussions with colleagues suggest that there are some systematic differences between fluent and less fluent writers that may be instructive. We organized some systematic discussion of this in my department a while ago and something very interesting emerged. A number of colleagues said that they were working on three or four different papers at a time, but these were the people who take a very long time to get any piece actually finished or submitted. By contrast, the fluent writers – including me – generally write one paper at a time. The only exception I make to this is when a journal accepts a paper subject to revisions. When this happens, I give total priority to revising the paper, even if it means suspending work on another paper in progress. By this means, I ensure the speediest possible publication of my work. Slow writers, by contrast, often put off working on such revisions for months, to the point where they may forget important details of the original work or where the editors lose patience and treat their eventual resubmission as a new paper. This, by the way, is a minor disaster, as it means the paper going to new referees who express new views and ask for further revisions, starting the whole process over again.

The 'one paper at a time' strategy is beneficial for me, as I like (mentally) to plan a paper as a whole and to feel its overall structure and story. Ideally, you would start writing the paper and keep on – uninterrupted by other tasks – until you finished, but only research students and those on sabbatical leave get this luxury. Multi-tasking time management is obligatory for academics in the modern environment. None of us likes it, but we have to live with it. Writing is perhaps the most vulnerable activity, so it is certainly a good idea to organize your time with this in mind. If your timetable gives you a clear day, then it is best to set this aside for writing and fit research meetings around busy days between classes and administrative meetings. Trying to fit writing into odd slots is not going to work because the process requires total focus and concentration.

The environment in which you write is important also. I believe that it is helpful for most people to do their writing in the same place. The choice will normally be between your office at work or your study at home. Some people like to do all their work in the office, keeping home entirely as a place of relaxation. Others, especially those who have young children in the house, find working at home impractical. For many years, I have done all my serious writing at home (children grown up and gone), as it is relatively free from interruptions. I just added 'relatively', as I received an infuriating unsolicited commercial phone call in the middle of writing that last sentence! You also need to be fairly relaxed and stress-free to write well. This is another reason to my mind for writing at home, or at least in a consistent and controlled environment. The work environment is not just busy and subject to interruptions but associated with tensions inappropriate to the writing mindset. Some academics like to work in internet cafés, where social facilitation and a regular supply of coffee apparently assist. I have never tried this myself but it is finding the right environment for you that is important.

One of the key elements in writing is to consider your audience and to empathise with their state of knowledge and their personal motivations. Normally, I write for a specialist audience such as reasoning researchers. This makes the task of audience empathy rather easy, as they are people like me with similar knowledge and interests. However, I still have to make allowance for the fact that they are unlikely to know as much about the specific topic as I do and nor can I take their interest entirely for granted. I still need to motivate their reading by the way in which I introduce the paper and set the context. A more complex problem was faced when writing a recent book on conditional statements (Evans & Over, 2004). The book was written in collaboration with a philosopher and brought together both philosophical and psychological work on the subject. It followed that the potential readership would be from both disciplines. I discussed this with my co-author, David Over, and we agreed to write the philosophy on the assumption that a psychologist was reading it, and similarly to write the psychology for the philosophers. Of course, we hope that everyone will read the entire book, but this was nevertheless a good mental discipline. Almost no-one will complain that you have written something that is too *easy* to read!

In writing (and revising) the current book, I have faced a quite different challenge, as my target audience consists of research students and early career researchers whose interests may fall in any area of psychology or even in other related disciplines. One approach that I have taken is to recall my own experience at the same career stage and to imagine those things that I needed to know, but did not at the time. Another is to consider what conventional research methods textbooks cover and what they leave out to provide a more objective basis for assessing the state of knowledge of my target audience. I also quite often imagine a particular individual reading my work as I am writing it. In this case, I see an early career researcher who is a good scientist but also motivated to have a successful career.

I assume that they are not working in cognitive psychology (the same principle as writing the psychology for the philosophers). I assume that they have been trained to a reasonable level of competence in statistics and research methods and so on. I ask myself frequently: is this reader still with me, have I kept their interest, was that section too technical or the other too simple?

Something else that good writers seem to have in common is that they enjoy reading themselves! Even though I engage in less re-drafting than most writers, I regard the reading back of my own text as a critical part of the process. In fact, I usually read my drafts several times both immediately on completion of a session of writing (if time permits) and also after a delay. This process is always accompanied by my imaginary reader from my target audience. I have noticed that some people do not much enjoy reading their own drafts (and if they do not, who will?). A former PhD student once gave me a draft thesis chapter to read, admittedly produced under time pressure, that was really poorly expressed and structured by his own quite decent standards. 'Have you read this?' I demanded to know. He admitted that he had not.

In Chapter 1, where I discussed the preparation of a literature review, I emphasised the importance of structuring the material. This actually applies to all forms of writing. As I have said several times, you must tell a story.[1] First, engage the reader's interest, then develop the plot and finally draw the whole piece to a satisfying conclusion. The empirical journal article provides the greatest challenge but is also the single most important form of publication in your scientific career. It is this that we now consider in detail.

Writing journal articles

Journals sometimes publish review or theoretical papers, but the empirical paper is the major form that you will need to master, and easily the most difficult to write. This is the means by which you will publish most of your research work and your success in this will determine pretty much everything else in your research career. Success means publishing most of the work you conduct in the best possible journal and in the most readable and accessible style. This, in turn, will determine reputation and citation, research funding, promotions and so on. It is worth emphasising these consequences in order to motivate the acquisition of this most difficult and complex skill.

Study the structure of a journal article, shown in Figure 7.1. Note the symmetrical structure in which you commence at the most general level at the start of the Introduction and return to it at the end of the General Discussion. Note also the coherence of the whole. The Introduction sets a general and then specific context for the empirical work, while the General Discussion discusses the findings in precisely that context. For those familiar with musical forms, this is analogous to sonata form: theme, development, recapitulation, coda. Indeed, like a sonata, or symphonic movement, the article needs an aesthetically pleasing balance in its construction. It is a story with a beginning, a middle and an end. However, do not expect that the story told will bear much resemblance to the actual story of how

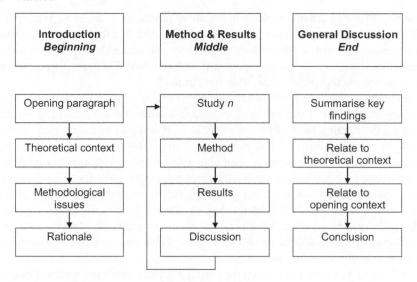

Figure 7.1 Structure of an empirical journal article

the research happened. It is not the purpose of a journal article to reflect the history of the research but rather to contribute knowledge to the discipline.

When you write for a journal, the research must be your own original work and should not have been previously published or submitted in parallel to another journal. You must not report predictions that were not actually made in advance, which is dishonest and affects the statistical inferences that you can make. However – and this may be surprising to newcomers – you may well be required to frame your research in terms that were not considered when the work was originally designed. This may seem deceitful, but journal editors not only permit this, they often *require* it. For example, your paper is submitted in ignorance (due to all too common failures of scholarship) of some relevant published work in the field that is identified by the referees. The editor may ask you to introduce the work in terms of these missing studies and build this into the rationale. Sometimes, you frame the submitted draft in terms of a particular theoretical approach but the editor asks you to drop this and use another. The theoretical framing and rationale are thus mutable and the only ethical rule is that you do not report predictions that were not actually made in advance.

Why does the story told in the journal bear so little relation to the actual story of the research? Actually, there is a very good reason. As I have said, the purpose of academic publication is to advance knowledge in the discipline, not in the head of the researcher who carried out the work. For example, the work may be original – to you – but if the findings are already known in publications unread by you, then the work will not be judged as advancing knowledge. Fortunately, this rarely happens, but it is common for authors to miss previous studies of relevance to their work that are pointed out by referees. In order to maximise contribution to knowledge, the journal editor will want your paper to be framed (in the

Introduction) by reference to the most up to date and relevant literature available. Hence, your article starts by summarising the current state of knowledge (of the field) and then proceeds to identify issues that need to be addressed. You report your empirical study and then show how it has advanced knowledge as a result. You can think of the three parts of the article in Bayesian terms (see Part Two) if you like: state prior belief; present new evidence; state revised belief.

In Chapters 1 and 3, I talked about the origin of ideas and the way in which people go about designing empirical studies. Reading and thinking about relevant literature is one source of ideas and studies but there are others, including intuitions, analogies from other fields of work, hypotheses formed from real-life observations and so on. You may be aware of some of the sources of your ideas and be unconscious of others, but it really does not matter when it comes to writing journal articles. The sonata form requires that you present your research according to a pre-set schema. Let us now consider the components (Figure 7.1) in more detail.

You may be surprised that I have identified the first component of the Introduction as the opening paragraph. However, such paragraphs are vital in my view to a well-written journal article. You should set it at the highest possible level of generality and one that reflects the real purpose of the research paradigm within which you are working. Psychological paradigms develop lives of their own, but they are always, in reality, *sub-goals* rather than ends in themselves. What they are sub-goals of is for you to identify in your opening paragraph. The paradigm may be hindsight bias, but the purpose is to understand the frailties of expert judgement that affect our daily lives; the paradigm may be dichotic listening, but the purpose is to understand the ability of human beings to switch their attention to the most significant of a mass of perceptual inputs that bombard our senses; the paradigm may be out-group stereotyping, but the purpose is to understand the impact of our cognitive style on our interactions in the social world – and so on. Never start your paper by direct reference to the paradigm as though it were an end in itself. Always remind the reader (and, indeed, yourself!) of the wider purpose. If you do not motivate them here, they may read no further than your first paragraph.

A couple of examples may help. In a paper published in the journal *Memory and Cognition* (Evans, Clibbens, Cattani, Harris, & Dennis, 2003), the responsibility fell to me as first author to write the main draft. The paper concerned a rather technical paradigm known as multicue probability learning, in which participants were run for many trials in which they were given multiple pieces of information with a probabilistic relationship to a criterion they were trying to learn by outcome feedback. The task – albeit artificial and computer based – ought to simulate the kind of real-life experiential learning that contributes to the development of expert judgements. Hence, I drafted the following opening paragraph:

> Expert judgment is often exercised in situations where multiple pieces of information are available and relevant in varying degrees. For example, a doctor may be attempting to diagnose the likely illness of a patient on the basis of symptoms, medical history, clinical examination, and results of imperfectly

diagnostic tests, as well as relevant demographic information such as gender, age, and occupation. To take another example, a personnel manager might be judging the suitability of a candidate for a position taking into account work experience, qualifications, interview performance, and results of psychometric tests. These are examples of what are termed multicue judgments, where the various dimensions of information are the cues, whose values have particular and distinct instantiations for individual cases.

Note that I have introduced here immediately the highest purpose of the paradigm: the understanding of real-world judgemental expertise. The experiments reported in this paper did not actually use expert participants or even realistic problem content, but the examples in the opening paragraph create a context to which the reader can easily relate and that provides the ultimate reference point for the rather technical and abstract research that was reported. The context is also one that *motivates* the reader to work through the technicalities to come. As the introduction to this article developed, the examples of the opening paragraph (doctor, personnel manager) were returned to in order to illustrate the significance of the methodology. The theoretical context (see Figure 7.1) was then developed around the notion of implicit and explicit thought processes that might underlie such judgements leading to a statement of the specific rationale for the experiments.

The second example is from a paper that I published with Simon Handley and David Over in the *Journal of Experimental Psychology: Learning, Memory & Cognition* (Evans, Handley, & Over, 2003). The paper was about conditional statements, introducing a new methodology and theory that would require considerable attention on the part of the reader to take in. We chose to start the paper by focussing why conditional statements are important and interesting in psychology:

> *If* is a two letter word that has troubled philosophers and psychologists alike. It cannot be escaped in the study of either reasoning or decision making. To infer a conclusion from premises in reasoning is to commit oneself to holding that the conclusion is true, or at least probably true, *if* the premises are true. To consider an option in decision making is to think what will happen, or at least probably happen, *if* the option is taken. *If* has properties quite unlike other propositional connectives such as *and*, *or* and *not*. In particular, a conditional statement invites the listener to focus on the possibility that its antecedent holds.

This paper inspired the production of a book published a year later (Evans & Over, 2004) with my shortest ever title: *If* (no subtitle). Our whole idea was to shift away from the idea that conditional sentences are a technical topic for study in philosophy and psychology and, towards that notion, that *if* is the cornerstone of hypothetical thinking, the basis for distinctively intelligent human thought. This short opening paragraph for our 2003 paper goes as far as it can towards achieving this aim.

Introductions must, of course, introduce. Their purpose is to set the contexts: general, theoretical and methodological, which support the rationale for the empirical reports to follow in the middle section. There are some important conventions to bear in mind here. A journal article introduction must be entirely *a priori*. It must be written as though the research had not yet been conducted and in no way anticipates results to come. This affects choice of literature to be reviewed. Studies discussed in the Introduction must only be those that support the *a priori* rationale for the paper (other studies whose relevance is indicated by the findings can be introduced in the Discussion). What is potentially confusing here is that we are talking *a priori* in the sense of the sonata form, since the framework for your introduction and some of the literature cited might be *post hoc*, so far as the actual design of the work is concerned, as discussed above. However, having decided on how to frame the rationale for the study, write as though it were elegantly deduced from a set of issues in the literature and then written before a single piece of data was collected! Do not review studies for the sake of it: they must be entirely relevant to the rationale or else should be omitted.

When writing method and results sections, there are some general rules to consider. Never make your paper more technical or more difficult to understand than it need be. Dense technical writing will not make you seem clever to the reader – it will simply reduce the probability that they will bother to read it thoroughly and understand your research. A more general principle is never to make your reader work harder than necessary. This means also that you should report your results selectively, dropping or de-emphasising findings that are of marginal interest or relevance to the story. Intriguing sub-plots are allowable; dull distractions are not. You should also give some thought and attention to the clearest ways of presenting your information. Do not burden your reader with unnecessarily complex tables with excessive significant figures. Two figure percentages are good, as are clear and simple diagrams and figures.

In a multiple experiment or multiple study paper, you may require intermediate discussions and linking sections. These are also part of the general story schema, reminding the reader of the higher level framing and rationale that underlies the study as a whole. However, I favour keeping such sections fairly short and dealing only with specific issues. So far as possible, defer consideration of deeper and more significant issues to the General Discussion. This last section is absolutely vital, for it is where the research ultimately delivers its purpose: the advancement of knowledge. As I indicated early in this book, I do not consider research to be the collection of empirical facts but rather the advancement of theory and understanding. Your empirical results do not do this on their own: they require interpretation and contextualisation, which is the purpose of the General Discussion.

Consider again the story grammar shown in Figure 7.1. The General Discussion should return through the levels and always link directly with the context provided by the Introduction. It is helpful to start with a summary of salient findings, which you can do here in a manner that is more selective and less technical than in the formal reports of the Results sections that preceded it. The main purpose of this section is then to interpret these findings in the context provided by the Introduction.

In writing the former section, you reviewed relevant studies and issues and showed that a certain problem needed to be addressed in order to support the rationale for your study. Now you return to these same issues, enlightened by the new findings reported, and conduct a public Bayesian revision process. Your results may have provided a novel empirical finding that you can emphasise. They may have repeated a previous finding with improved methodology; they may have supported or failed to support a key prediction that would favour one theory over another and so on.

Since your paper started with a reference to the loftier purposes of the research in the opening paragraph, try to return to this broader context towards the end of the Discussion. Do not simply interpret your findings in terms of the specific theoretical and methodological context, but return to the starting point. What does the revised belief imply for expert judges, social behaviour or attention in the real world? Finally, your paper must have a concluding paragraph – a take home message – which can be harder to write than the opening paragraph. Try to avoid the lame 'further research is needed'. To be sure, one purpose of your Discussion is to point to issues that need to be addressed further, but it is a truism that further research is (always) needed. It is better to summarise in clear and general terms what we know as a result of this research work that we did not know before.

Editors and referees

When you submit your article to a journal, it will be subject to peer review, which may be a more or less difficult process to deal with depending on the status of the journal and its rejection rate. Broadly, there are three main kinds of decision that you may get. The paper may be accepted at the first attempt, usually subject to revisions to be approved by the editor. This will be a relatively rare experience to start with. Second, you may be encouraged to revise and resubmit the paper for further consideration. This is what you will realistically be hoping for. Finally, the paper may simply be rejected. You will get papers rejected. I do; everyone does. The only people who do not get articles rejected are those not submitting them. However, with experience, you will learn to minimise the rejection rate.

The reasons that a paper may be rejected are the following:

(a) The paper is unsuitable for the journal. For example, it falls outside of its topics of interest, or you have sent a review article to a journal specialising in empirical papers and so on. These rejections can be avoided by proper study of editorial policy and the Advice to Authors that appears on the flap of most journals and can be accessed online.
(b) The paper is methodologically flawed.
(c) The paper is methodologically sound but makes an insufficient contribution to knowledge. The findings are not novel or have insufficient theoretical significance.
(d) The paper is badly written and too difficult to understand.
(e) The research is poorly motivated, lacking a clear theoretical foundation or practical objective.

Rejections can be painful, especially early in your career or if you get several in a row. However, it is important to learn from them. The peer-review process is generally fair – though subject to random variation – and if your papers are consistently being rejected, then you are definitely doing something wrong. If, say, your methodology or writing style is regularly being criticised, then look at the criticisms and try to remedy the problem. Most rejections, however, come into category c above – the work is judged not to advance knowledge sufficiently for the journal to publish it. Avoiding this kind of rejection is partly a matter of developing theoretical knowledge and the skill of designing empirical studies that address salient issues effectively. However, it is also partly a matter of skill in *writing* journal articles.[2]

In the previous section, I placed great emphasis in the fact that psychological research needs not simply to be reported but also interpreted and that this interpretation is part of the advancement of knowledge. It follows from this analysis that what journals will perceive as a contribution to knowledge will depend not just upon what the research is but also upon how it is written up. A more skilled writer could get a paper accepted on exactly the same empirical study that would result in a rejection on grounds of insufficient knowledge development in the hands of a less skilled writer. An inexperienced writer may get their paper published in a journal of moderate status, whereas with more experience and different writing, the paper could have been published in a higher standing journal. Having said this, you cannot simply regard writing as a content-free skill. The ability to interpret research effectively also depends on the level of scholarship and depth of understanding that has been acquired (see Chapter 1). Hence, it is to be expected that young researchers will publish in more modest journals than those they will target as experience and knowledge grows.

If a paper is clearly rejected, then accept the decision and resist any temptation to write an angry letter or email to the editor. Did you ever see a referee rescind a sending-off because the soccer player in question argued with his decision? Journal editors are the same. They will not change their minds and since editors are generally quite powerful people, you could do your career some damage by telling them what you think of their decision. There is, however, nothing wrong with politely seeking clarification of a decision through an email exchange, provided that you make it clear that you are accepting and not contesting the decision.

The critical category of response is that which encourages – or at least permits – resubmission of a revised paper. You will normally receive an editorial letter plus two to four reviews from expert commentators, who may or may not choose to reveal their identities (many do so nowadays). The most important of these is the editorial letter because the editor has the ultimate power to decide the fate of your paper. Sometimes, the tone of the editor's letter is more positive than that of the referees and sometimes it is more negative. In such cases, the editor's tone is the one you should attend to in deciding whether to resubmit your paper or try your luck with another journal. If it is negatively inclined, but leaving the door open, then study carefully what the editor's concerns are and also try to evaluate his/her expertise in your field of study. Sometimes, editors are not very expert on the subject and may have misinterpreted something. In such cases, do not be afraid to engage the editor in an email exchange to try to sort out the problem. I quite often negotiate the terms of a resubmission by such exchanges with an editor rather than just relying on the editorial letter.

Referees' reports usually include a fairly clear (sometimes implied) recommendation to the editor as to the decision to be made – accept, resubmit or reject. Of course, the set of reports may make conflicting recommendations, which is where editors' judgements come in. Where reviews are negative, it is important to analyse them carefully. If the response makes you angry, then return to the reviews after you have cooled down and try to view them objectively. Here are some possible causes of a negative review:

(a) Your paper has significant flaws in methodology, scholarship or writing, which the referee has rightly and correctly identified.
(b) The referee is incompetent or lazy and has not bothered to read your paper properly and assess it accurately.
(c) The referee is biased by her or his own theoretical and published position and hostile to your work, which in some way criticises or threatens to undermine their own.

In my experience, reviews of type (b) are fairly rare: most referees are competent and professional and will read your paper quite carefully. If your analyses show a high frequency of (b) reports, then I suggest that the problem is in your own mind. The frequency of (a) depends, of course, on your skill as a researcher. However, you will make mistakes and a cool reading of fair criticism will help you to learn from them. If you cannot take criticism, then you should seek a different career. Academics are subjected to critical comment throughout their careers. Even biased reports may contain some fair criticisms, which you must be able to recognise and respond to if you are to make progress. Reviews of type (c) are actually quite common, however. Since your reviewers will be experts on the subject, they will also have a stake in it. They will not like it if you have omitted reference to their own papers on the subject and may also respond defensively if you have criticised their work. Some referees are a lot more partial than others. One of the responsibilities of a journal editor is to learn which referees are more interested in their own work than that of the submitted authors. In the worst cases, they can stop using them or at least make allowance for their bias in reading their reviews.

Biased reviews may sometimes lead to outright rejection of a paper. This is most likely to happen to you when you are in early career and have not yet established a name of your own. It is very frustrating, but the only recourse is to submit to another journal and hope to avoid the same referee receiving your paper again. Some will decline the opportunity to review your paper for a second journal but there is no clear convention on this. If – for whatever reason – a paper is rejected and you decide to submit to another journal, it is important always to study the reviews for fair and constructive criticism first and make changes accordingly. First, this is likely to improve your paper and its chances of success with the new journal. Second, there is a larger chance than you may think that at least one of the reviewers will have seen the paper before. If they realise that you have submitted elsewhere without making any changes, then you could well be seen as arrogant, which needless to say will not help your chances of securing publication.

As an author, it may help to consider the perspectives of referees and editors, which, in any case, you should progress to be yourself in due course. The editor's role depends on the nature of the journal and its rejection rate. With middle ranked journals, the editor is normally quite keen to find papers of publishable quality to ensure continuation of the journal. Hence, they will not wish to reject your paper outright unless they think it is beyond redemption. Some high ranked journals, however, have submission rates way in excess of what can be published and will have to reject many competent papers. In such cases, the editors will look for an exceptional reason to publish rather than needing a reason to reject your paper. Referees are researchers who give up their time – unpaid – to help evaluate manuscripts of other authors. They may do this out of a sense of professional obligation because the system depends upon peer review; they may also do it because it keeps them abreast of current developments in the field; a motive for a small minority would seem to be advancement of their own careers by insisting on references to their own work being added or by rejecting authors whose work undermines their own.

Dealing with bias in referees is a tricky issue. Some experienced authors may request of an editor not to use a particular referee that they have had problems with in the past. If you do this (which I rarely do), do not overdo it. I once had a very experienced author send me a list of about 16 names of people he did not want to review his paper! This is unreasonable and may indicate lack of confidence in the paper (in this actual case, I used three referees who were not among the 16 and they all recommended rejection). If some of your referees are biased against you, that is the way of the world; if most are, then you are suffering from paranoia! If you think you have a real problem with a particular individual, then a private and polite email exchange with the editor might be the best approach. Any formal correspondence between you and the editor will have to be copied to the referees, which constrains what can be said on both sides. You should normally be able to trust editors in this regard.

When you resubmit a paper, make life as easy as possible for the editor. The best approach is to accompany your resubmission with a letter in which you detail responses to every single point in the editor's letter and the referees' reports. Where criticisms were fair and sound, you have made changes in the manuscript to which you can refer in the letter. In a minority of cases, however, you can refrain from making changes and instead present an argument or rebuttal in your resubmission letter. The referee may have made a statistical or methodological criticism that was unsound. Politely point this out in your letter, explaining misunderstandings or giving reference to a technique with which the referee may be unfamiliar. If you can show that you have dealt with all the points in this way, it is very difficult for editors to reject your resubmission. In any case, they will already be looking favourably upon you, as you have made their job so easy.

When revising a paper, there may be some major issues that you need to deal with first, after discussion with your collaborators. However, when you get to the detail, the revision can be a complex and demanding process, as you have to respond to a number of different points from several referees as well as the editor. The way I do it these days is as follows. I go through the editor's letter and each of the referees' reports, numbering each distinct point, assigning a prefix A, B, C, etc to each reviewer. Then I go through each report in turn and mark the manuscript with the point that applies in the appropriate margin location. Having done this, I then end up with a manuscript that is marked in the correct sequence: A1, A2, B1, C1, A3, B2, B3, etc. Then I revise the paper from start to finish (the best way to preserve the story structure), referring to each report as required whenever a margin note is encountered. I also keep open another text file that will contain my eventual resubmission letter to the editor, so that I can note what I have done with each point as I proceed through. However, I enter these notes for the editor into the section corresponding to each referee. At the end of this process, I have processed all points in order of the manuscript but end up with a letter to the editor in which I respond to each report in turn. It took me a long time to figure out this way of doing things, so I hope it works as well for you as it does for me.

Writing brief articles

There has been an increasing trend in psychology towards the publication of short articles, usually between 3,000 and 5,000 words in length. Some journals specialise in them (e.g. *Psychological Science*) and many others now carry a section that allows publication of such short papers. An advantage is that publication lags are usually a lot shorter. The status attached to these papers and the potential for citation is just as high as with a full article. Sometimes, citations will actually be higher because more people will take the time to read something short and simple.

The first issue is to consider whether the research you have to report is suitable for a brief article. Many that are submitted are rejected by the major journals because researchers misinterpret the purpose of these publications. They are *not* there to dump small studies and minor findings into print, as a lot of authors seem to assume. Their purpose is to provide rapid publication of findings that hold exceptional interest and that can also be effectively communicated in a small number of words. I must admit to having made the mistake more than once of writing a brief article on the grounds that the study I had to report was small (e.g. a student project) or on the lazy assumption that I could get away with writing a shorter paper. These submissions all ended badly, as you might expect. In fact, on several occasions, I have had a short article rejected with a recommendation to submit a full article reporting the same research, which was later accepted. I have also published a couple of short articles that attracted better than usual citations for my full articles but only when they met the criteria described below.

The real basis for writing a short paper is that you have an unusually simple and clear story to tell. Normally, the paper should focus on a single issue and single finding. Short papers often report a single study, but more than one experiment is possible if they can be briefly described. The method needs to be sound, the finding clear cut and its impact original and stimulating. It also needs to be something that will stand on its own without a great deal of space being needed for building the rationale or interpreting the findings. That is why research submitted in the brief article format is often more likely to succeed if written as a conventional full article.

Writing short articles is an art form. It is not easier than writing a full article; in fact, it is more difficult. It creates similar problems as that of trying to communicate your research in one of those wretched 15 minute conference slots. So do not think of it as a soft option or a first recourse. Only occasionally will you have some research findings that really suit publication in this format. The findings need to be simpler to describe and report than usual, but no less interesting and important. If anything, they need to have more immediate significance.

Opinion pieces

Something relatively new is the ability to write opinion pieces as journal articles. There are journals that will accept articles of this kind (e.g. *Trends in Cognitive Sciences*) and some that specialise in them (e.g. *Perspectives on Psychological Science*). Although worth a mention, these are generally not available as an outlet for early career researchers, except perhaps in collaboration with experienced authors. They are mostly for mature or even late career researchers who have had many years to acquire knowledge and wisdom and can hence offer informed opinions. Also, readers are more likely to respect the opinions of those who have established over many years that they have made a major contribution to the topic and have something to say about it.

Although I do not recommend that early career researchers try to write opinion pieces, I do recommend that they read them! The conventional journal article, as explained above, is a highly constrained and, in many ways, artificial form. Opinion pieces allow experienced researchers to give their views on how their research fields are progressing, what is good and not so good in current practice, how things might be expected to develop in the future and so on. These are the kinds of things you could previously only have learned by actually meeting and talking with the individual concerned. They could prove very valuable to early career researchers trying to understand a research field and anticipate how it is likely to develop.

Writing books and book chapters

While writing empirical journal articles is the bread and butter, there are other opportunities for academic writing to be considered. For example, you may be asked to contribute a chapter to an edited book. Some such books are effectively multi-authored student textbooks but others are aimed at research fields. I must say that I am not at all keen on the practice of publishing reports of empirical studies in book chapters, as these should always be subject to the rigours of peer review. However, book chapters may serve a useful purpose when they give the opportunity to write reviews or essays. If you have ambitions eventually to write for such lofty journals as *Psychological Bulletin* or *Psychological Review*, then an edited book chapter may provide a useful opportunity for practicing the essential skills.

I discussed the way in which to construct a literature review in some detail in Chapter 1 and will not repeat that advice here. An *essay* is a form that really only the book chapter (or a short book) permits. Essays are fun because you set out to argue some point or position and illustrate it with reference to the academic literature, your own published and unpublished work, theoretical conjecture or some combination of the above. You will get a freedom that the conventional journal article does not allow. An essay can be quite similar to an opinion piece, but unlike the journal format, a book chapter of this kind may be written in early career. Most of my book chapters are essays, and I started writing them early in my career from 1980 onwards.

Although books are subject to some form of peer review, it is usually a great deal less rigorous and constraining than the refereeing system employed by scientific journals. Typically, the book editor will provide some comments and suggestions and perhaps send your chapter to one of the other contributors for comment. Your chapter will generally have been invited and so is not likely to be left out unless it is really bad or way behind the agreed schedule. This is both an advantage and a weakness. The nice part is that you have relative freedom of expression without referees breathing down your neck and enforcing changes you do not really want to make. The drawback is that without rigorous peer review, quality may suffer, which is particularly problematic for empirical reports. For similar reasons, book chapters generally attract fewer readers and fewer citations than journal articles and carry less weight on your CV.

Books fall into several categories. There are undergraduate textbooks, which have more to do with teaching than research: they *might* make you some money, but they will not attract many citations in the research literature and will do little for your reputation as a researcher. There are specialised texts, which are essentially reviews aimed at research fields. If successful, such books may attract many citations, but they will definitely not make you very much money! Generally speaking, academic books return royalties at a very miserly hourly rate when you consider how long it takes to write them. The reason to write them is because you have something to say that requires the length and freedom of a book to express. Writing the right books at the right times can also have a very clear career benefit, for reasons I discuss below.

For the serious researcher who likes to think deeply about their research field, there is another kind of book to consider. There are extended essays and monographs in which you can present your theoretical ideas at length and with freedom from detailed technical reviewers (although most publishers have some kind of review done, if only to suggest improvements). I have published several books that fall into this category (Evans, 1989; 2007a; 2010; Evans & Over, 1996; 2004) and have no reason to regret writing any of them. In fact, these books probably helped my career more than any of my journal articles, as all were well read and cited by my target audience. Of course, you cannot write one unless you have something to say that will sustain writing at this length.

The criteria are somewhat different for a book, as academic publishers are busi-nesses. They are not necessarily looking for money-spinners but they will need convincing that there is a reasonable market so that the book will at least cover its costs of production. Good academic publishers are also concerned about their reputations and will have proposals carefully reviewed, especially if presented by young researchers of limited reputation. They tend, on the other hand, to compete fairly hard to sign up books by established names.

Publication strategy: political and career issues

Some people have more research time than others by virtue of their working and/ or domestic situation. Some work more quickly or write more fluently. Some people are only comfortable writing up a set of empirical studies, while others have a lot of ideas that they want to express. All of these considerations will fac-tor into whether you will be a book writer as well as an author of journal articles. You can make a good career limiting yourself to articles. Generally speaking, you can make a better one if you can include some books as well, especially if people find them interesting and readable. Books can achieve surprisingly high citation rates in academic journals, often well beyond that achievable by journal articles. They can build reputations and, in some cases, make reputations from nothing.

If you are not a particularly fluent or prolific writer or if your research time is limited, the focus should be on journal articles. Even if you are a good writer, you may need 10 years or so experience in your research field before there is a book in you worth writing. If you are a slow writer, then you should be wary of books and book chapters, as your production of journal articles will suffer as a conse-quence. However, the view that I commonly hear that really *only* journal articles count is greatly exaggerated. Some book chapters are major sources of inspiration in psychology and cited with huge frequency: Wason's (1966) chapter introduc-ing the four card selection task and Baddeley and Hitch's (1974) seminal paper on working memory are two examples that leap readily to mind. (Admittedly, these were established names at the time of writing and it is doubtful whether a young researcher could have major impact with a book chapter.) Each launched a major research paradigm that thrives to the current day. Books, of course, can be enormously influential and can reach much wider audiences than journal articles. I think most researchers' reputations can be substantially increased if they publish the right kind of book at the right time. Hence, there may come a point when writ-ing a book will most definitely be the right thing to do.

In considering your strategy for publication, you need to be patient and to do the right things for the career stage in which you find yourself. The biggest issue really is which journals you should try to publish your work in. I would ignore low ranked journals – a soft but rather pointless option – and concentrate early on in establishing yourself in middle ranking journals that carry international standing but are not as difficult to get into as those at the top of tree. Having thus provided a good basis of solid empirical publication and developed skill and experience in both research and publication, you should then try to step up a level and submit some of your papers to really ambitious journals. If they are rejected, learn from the experience and still publish the work in the journals you know well. Consider also the possibility of writing review or theoretical articles, perhaps taking on a few book chapters to practise the skill before targeting prestigious journals. There may come a point – probably at least 10 years subsequent to your PhD completion – when you might feel ready to take on a monograph or essay-style book.

For much of my early career, I misunderstood the role of high profile publishing. As usual, I could never really come to terms with the prestige aspect of university politics, or more crudely put, the pervasiveness of academic snobbery. My naïve perspective was that if you published good work in accessible places, it should get the impact it deserved. The address on the letterhead or the standing of the journal should make no difference. Sadly, nothing could be further from the truth in the real politics of academia. Prestige and status are overpowering elements in the system. Here are some examples from my own experience:

(a) I once said to an American psychologist, 'Are there not a number of universities in [unnamed large city]?' 'Yes', she said,' but only two that matter'.
(b) I lost count of the number of people who said to me in the past, 'Why do you work at Plymouth when you could work anywhere you want?' (This was before my department built what is now a strong reputation as a good place to work.)
(c) On a sabbatical visit to the USA, I mentioned that some of my papers were relevant to a PhD thesis and was told, 'My supervisor said I could ignore those because they were published in European journals'.

I could give many more examples, but you get the drift. Much more insidious is the implicit influence of prestige. Here is the simple fact of the matter. If you publish in a prestigious journal, then people will assume that the paper is important, and will be more likely to read it, cite it and take it into account in their own work. Papers in middle ranked journals will have a lesser impact and below that may be ignored altogether. This is why there is so much attention to impact factors of journals and why many authors will study these stats before deciding whether to submit to a given journal. Impact factors generally measure the average rate at which papers in the journal get cited, skewed towards recent publications.

There is something very misleading about impact factors of which you need to be aware. Review and theoretical articles will generally receive a lot more citations than experimental papers. This is because they have a wider field of influence. Many authors might be interested generally in a topic like, say, working memory and will be likely to cite a recent review article in a prestigious journal like *Annual Review of Psychology* or *Psychological Bulletin.* Because empirical reports are much more specific and limited in scope, they will tend to be cited only if relevant to details of the particular piece of research an author is citing. Hence, when you compare citation rates between review and experimental journals, you are not comparing like with like. Even empirical journals of very good reputation that carry much prestige will have citation rates a lot lower than those that allow review or theoretical papers. If you are concerned about the impact of your own papers, of course, you might consider including some papers of the review/theoretical type to boost your own personal citation ratings.

Online and open-access publishing

There is currently a sea-change going on in academic publishing. It is evident that from a technical point of view, journal publication needs to be an online business rather than a book business. No researcher these days wants to receive a bound copy of a journal publishing articles on a variety of topics, most of which they will never read. We all now have electronic library subscriptions through our universities. What modern information systems allow us to do is to make electronic searches for articles on the topics that interest us, no matter where they are published. When we find them, we want to pull them down as pdf files and print them (or not) for our own use. Provided our library has a subscription, that is what the vast majority of us do. So why does the library need to be cluttered up with all these book-like bound volumes? In fact, why do journals need to publish in paper form at all?

The answer really is that we are in transition from one form of publishing to another, and the academic publishers have not caught up, or are resisting for reasons of commercial interest. They want to restrict access to their journals so that readers (or, usually, university libraries) have to pay high fees to be able to read them. Only with appropriate and expensive licences can such articles be downloaded as pdf files. There has been an orchestrated international campaign against academic publishers (some in particular) across all sciences in recent years for this practice, with publishers being vilified as greedy and holding back progress. I will explain some of the issues. The main complaints are first that publishers rarely pay fees to editors and never to referees or authors, so that the great majority of their product is provided 'free' by the academic world. In reality, it is provided at the cost of the universities who pay the academic salaries. These publishers then sell the same product back to the same universities for their libraries at what are sometimes considered to be extortionate rates. Not only that, but they hold up research progress because the naturally fast exchange of information that the internet allows is being blocked. In principle, a research article that has passed peer review could be instantly and universally accessible. In practice, prestigious journals will typically hold up publication for months and then only distribute articles to those who pay their licence fees.

While this debate has been raging, online-only journals have been springing up all over the place, often described as 'open access'. These journals do, in fact, make research instantly and freely available once papers are accepted for publication. But, of course, there is a snag. They charge money to publish these articles and surprisingly large amounts of it. One thing has not changed. Universities still provide the product for free through the writing and editorial work of their academic staff, and still have to pay large amounts of money for its publication. The only difference really is that the burden of financing publication is shifted from the readers (e.g. libraries) to the writers. It still seems to be the case that academic publishers are making a nice profit out of academia.

The situation is rather more complex than I have described because most conventional journals these days publish electronically as well as on paper and will offer an open-access option when you publish. This means that you will not have to licence the journal to access that particular article. But, of course, they also charge you a large fee for the privilege. In the UK, the government response to criticism of academic publishers has been to require publication of all publically funded research to be openly accessible by one means or another. To have done this unilaterally seems unwise to me. Academics still have to publish in the most prestigious journals possible because of all the other pressures on them and additional costs may be incurred to make these open access. (Many publishers will allow 'green' open access where the final manuscript is posted to an accessible website.) The total pot of public research funding has not been increased accordingly, so the effect of the policy might be to reduce the amount of money actually available for supporting research work.

There is another quite different concern that I have about the advent of online publishing. When journals are published on paper, there is a natural physical and cost limitation on the size of the volumes and, hence, the number of papers that can be published. Journals have acquired prestige when they are sent far more papers than they could publish and have accordingly restricted acceptance to those seen as the best. When this works well, it protects the academic community against publication of papers of little merit. Either they are not published at all, or in sources known to accommodate mundane work. The problem with online publishing is that electronic storage is cheap, as is access through the internet: much, much cheaper than printing and distributing paper volumes. So, in principle, much larger numbers of papers can be published. If the new style academic publishers have the same profit motives being attributed to the traditional ones (and I see no evidence to the contrary), they will want as many papers as possible accepted.

The evidence this is happening is obvious. New online journals are starting up every week, or so it seems, and bombarding academic email addresses (easy to obtain) with promotional materials. Those open-access journals that have already established themselves are promising 'light-touch' reviewing and limiting the power of editors greatly compared with the traditional role. I find the implications of all this quite scary. It suggests that both the quantity and quality control of the traditional system is endangered. There is already far too much work published for anyone to comprehend. The idea that there will be a great deal more with less rigorous peer review and quality control is not attractive at all.

These issues arise from free market economics working through such a cheap and easy means of publication and distribution as the internet. If we could somehow regulate academic publishing on an international basis, then we could combine the benefits of the traditional and new systems. We would have open electronic access, while also ensuring high quality control and limiting excessive profits for academic publishers. But I do not see how that can possibly happen and in the meantime, the whole academic publishing business is a chaotic mess.

I find this an area in which it is difficult to give clear advice to early career researchers because the whole system is in such a state of flux. I really do not know what academic publishing will look like in 10 or 20 years' time from now, although I expect that electronic and open access in some form will become the norm. I have tried to explain the merits and demerits of this compared with traditional academic publishing. As to which kinds of journals you should target, you will need to keep abreast of developments, especially the requirements of funding bodies and the policy of your own university.

KEY POINTS

- Writing is the most important of all the research skills to master. Without published output, you are invisible. Publication in higher ranked journals is determined by quality of writing, not just the research reported
- Although much more limited in impact, skill in oral presentations also has a useful role for young researchers, who can impress influential researchers with cogent and engaging presentations at conferences, workshops and departmental colloquia
- Good writing requires a consistent and conducive environment. According to the personality and circumstances of the individual, this could be in a university office, a home study or even an internet café
- Empirical journal articles are the most difficult form to master. It is essential that they engage the interest of readers from the start and tell a well-structured story. Skill in writing these is essential for a good career. Brief articles are even harder to write and should only be attempted where one has exceptionally clear and simple findings that can be presented in a limited number of words
- Articles are most usually accepted only after substantial revisions and, sometimes, additional empirical work. One needs a systematic approach to taking account of multiple complex commentaries and always an eye for the main decision maker – the journal editor
- Dealing with rejection of journal articles is a requirement of all active researchers. It is important to study reviews and learn from criticism. Experience will minimise but not eliminate such rejections
- Review article and opinion pieces in journals tend to come with maturity and can achieve high impact and influence. For a young researcher, accepting invitations to write chapters in books can provide an opportunity to practise relevant skills

- Writing undergraduate textbooks will not advance a research career, but other kinds of books may. Specialised texts that review research fields can be influential, as can book length essays or monographs that develop theoretical ideas at length. Book writing is for fluent writers, as journal output must be maintained
- Academic journal publishing is in the process of changing from one based on bound volumes to a largely electronic system. There are current complex political issues concerning academic publishers, open access and quality control

PART TWO

PHILOSOPHY AND PSYCHOLOGY OF RESEARCH

8   Hypothesis testing and reasoning

The provision of practical advice is the major objective of this book but not the only one. Being a good researcher also requires understanding of the process of science. In this chapter and the next, I focus on both philosophical and psychological issues that bear on this process. Philosophical issues concern the nature of science, what constitutes scientific knowledge and how we may best obtain it. I discuss the nature of logical or deductive inference in this chapter and statistical inference in the next. I will contrast two broad philosophies of science as Popperian and Bayesian and express a clear preference for the latter. Psychological issues concern what is known about thinking and reasoning and how it may influence scientific thinking, especially through the operation of cognitive biases. Understanding these should help you to avoid inferential errors that too commonly occur in psychological research.

Three kinds of reasoning

Science – all science – involves reasoning. We scientists do not merely collect facts. We devised theories and test them, we interpret facts, we provide explanations. None of these things are possible without inferences. However, the types of inference that scientists make fall into three clear categories that philosophers have distinguished *deductive*, *inductive* and *abductive*. Deductive inferences are those that follow the laws of logic. They are said to be truth persevering because if their assumptions are true, then their conclusions are bound to be true also. Any argument that does this is described as logically *valid*. However, a valid argument can have a false conclusion if one or more premise is false. Hence, we also use the term *sound* to describe a valid argument whose conclusions are true. Consider the following argument:

> All psychologists are scientists
> No psychologists are good writers
> Therefore, some scientists are not good writers

Is this argument valid? Yes. Is it sound? Probably not, unless you really believe the second premise. (There may be some who would believe the second premise but not the first!) Why is it valid? Here, we have to assume, like it or not, that the premises are true. There is a subset of scientists (psychologists) that are not good writers, so some of the scientists must be poor writers. But had the argument been written as follows:

> Some psychologists are scientists
> No psychologists are good writers
> Therefore, some good writers are not scientists

it would not be valid because we can find a *counterexample*: a state of affairs in which the premises hold but the conclusion does not. To understand this, study Figure 8.1. The premises take the form All B are A; No B are C; where A = scientists; B = psychologists and C = good writers. Surprisingly, there are three different situations consistent with these premises. In case 1, there is an overlap between C and A; in case 2, all of C is separate from A and in case 3, all of C is contained within A. In other words, it is possible that some, none or all of the good writers are scientists. Our first conclusion, 'Some scientists are not good writers', is valid because in all three cases, there are at least some scientists (those who are psychologists) who are not good writers. The second syllogism is invalid because the conclusion 'Some good writers are not scientists' has a counterexample in case 3. It is a logical possibility that scientists include all of the good writers as well as all of the psychologists, even though none of the psychologists can write well.

There are a number of formal methods, including mathematics, for ensuring and testing the validity of scientific arguments. I will not go into those here. The important point is that deductive reasoning is required to generate predictions and test hypotheses. However, this is not the only kind of reasoning that scientists need to do. The great drawback of logic is that you can learn nothing new by its use. All you can do is bring out conclusions implicit in what you already know and believe. Surely, an important role of science is to discover the laws of nature by observation? Indeed it is, but this involves inductive inference, which always

Premises: All B are A; No B are C

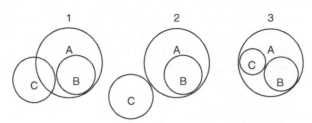

Conclusion: Some A are not C (valid)
Conclusion: Some C are not A (invalid)

Figure 8.1 Models of premises for the syllogism given in the text. In the example given in the text: A = scientists; B = psychologists; C = good writers

goes beyond the given information. Consider Newton's law of gravity, a thing of wonder in its simplicity and clarity:

> The gravitational attraction between two objects is proportional to the product of their masses and inversely proportional to the square of the distance between them.

This explains why gravity becomes very weak when we fly in a space ship. It also tells us that we exert the same gravitational attraction on the Earth as it does on us. So why do we fall on the Earth and not the other way around? For that we need another equation:

Force = mass × acceleration

Because our mass is so tiny compared with that of the Earth's, we are accelerated towards it rapidly, while its acceleration towards us is negligible. But how did we come to know these basic physical laws? Basically, by observation, *induction* and hypothesis testing. We observe the regularities in the behaviour of physical objects and try to generalise these into laws: an inductive inference. We can then test these laws *deductively* by applying them to new situations we have not observed before. But the induction of scientific laws is not a logically valid inference: the laws do not have to be true just because they are based on accurate observations, no matter how numerous. In fact, Newton's laws can be falsified under the special conditions of observation specified by relativity theory.

The third kind of inference that is important in science is *abduction*: reasoning to best explanation. Scientists are frequently called upon to provide explanations. Why are global temperatures increasing? Why did the AIDS epidemic spread so rapidly? What disease underlies a particular pattern of symptoms? Why was there a massive swing in voting option in that last 24 hours of an election campaign? Why is the world economy plunging into recession? These examples show that, for the general public, explanation is a lot of what we require of science. It helps us to understand, to learn and perhaps predict the future.

Like induction, abduction is not logically valid. (A broader definition of inductive inference includes abduction as a sub-type.) A classic type of abductive inference is medical diagnosis. We observe an effect (symptoms) and work backwards to the cause (disease). But diagnostic inferences are uncertain and usually lead to other medical tests to try to increase confidence. Global warming is occurring (air and sea temperatures have been steadily rising for many years) but what the public really want to know is what is causing it and can anything be done to prevent it. The greenhouse gas theory, with humans largely responsible, is the one that appears to hold the majority view among climate scientists, but it is certainly not the only explanation that has been seriously considered. Abductive explanations achieve at best a high degree of subjective belief or probability.

The importance of logic in classical philosophy of science

In my experience, few psychologists spend time reading philosophy of science and I imagine this is true of most empirical scientists. However, everyone who practises science *has* a philosophy, whether they know it or not. How do you test a theory? How do you know that you know something? If you are wrong, how can you find out? None of these questions can be answered by empirical methods. They are *a priori*, and hence philosophical questions.

Until comparatively recently, logic had a very good press in both philosophy and psychology. Philosophers loved deduction and worried a good deal about induction. Psychologists (e.g. Piaget) identified logical reasoning as the corner-stone of human intelligence and rationality. A field of psychological study of deductive reasoning grew up to test the ability of ordinary people to engage in logical argument. Unfortunately for the logicists, however, such studies have shown even intelligent adults to be quite poor at deductive reasoning, exhibiting many errors and biases (for reviews, see Evans, 2007a; Evans et al., 1993; Manktelow, 2012). For this reason, there has been a shift away from logic in the psychology of reasoning, as I discuss later.

It is not difficult to see the relevance of logic to scientific reasoning. A hypothesis is a proposition: it is an assertion about the state of the world that may be true or false. A theory is a set of related propositions that together have deductive consequences. The philosophical tradition known as logical positivism (Murzi, 2001) was built around the idea of treating scientific predictions and explanations as logical arguments. Consider the following argument:

> All metals expand when heated
> Allium is a metal
> Therefore, allium expands when heated

This is a deductively valid syllogism. The conclusion, however, is also a *prediction* if allium is a newly discovered metal that no-one has tried heating before. Hence, according to this view, scientific predictions have the form of valid arguments that seem very satisfactory. Suppose we already knew that allium expanded when heated. We could then use the same logical argument as a form of explanation. Allium expands when heated *because* it is a metal. All very neat – or so it seems. Applying this approach to psychological experiments is anything but simple, however, as I will show later.

A traditional difficulty in logical approaches to philosophy of science is the *problem of induction*. As already stated, a deductive argument has the great advantage that the conclusion is guaranteed to be true, if the premises are true. But it also has the massive disadvantage that you cannot learn anything new from it. Hence, we need inductive arguments such as:

> All the metals that I have ever tested expand when heated; therefore, all metals expand when heated

The conclusion here is more informative than the premise (the technical definition of semantic informativeness is that it rules out more possibilities; or if you prefer, is consistent with fewer states of the world). Inductive arguments have, in turn, the great disadvantage that they are not logically valid. Their conclusions are not guaranteed to be true. However many metals I test, there may be one that I have not encountered yet that will not expand. Many people live their lives seeing only white swans, but black ones nevertheless exist. The problem of induction is that scientific laws appear to be inductive generalisations. If the object of science is to develop such laws, then it is a discipline based upon faulty logic.

The proposed solution that is most familiar to psychologists is the falsificationism of the famous philosopher of science Karl Popper (1959; 1962). According to Popper, science is a process of conjectures and refutations. We construct theories and assert hypotheses that are then put to empirical test. These theories have to be falsifiable to count as scientific theories in the first place. He pointed out that although a universal claim can never logically be proved true, it can be refuted. Consider the syllogism about metals:

> All metals expand when heated
> Allium is a metal
> Therefore, allium expands when heated

Suppose that allium is a newly discovered metal that does indeed expand when heated. This does not prove the scientific law 'All metals expand when heated' to be true. In Popper-speak, it is *corroborated* by observation but not confirmed by it. Suppose instead that allium refused to expand. Then, Popper argues, we can refute our supposed law by valid deductive argument (known as Modus Tollens). A strictly universal law can brook no exceptions. If allium is a metal and does not expand, then the law is wrong.

Popper's philosophy of science is the one that most psychologists seem to have heard of and many regard it as gospel. If psychology is a science, then our theories must be constructed to be falsifiable and experiments must be designed that could demonstrate such falsity. According to this view of science, theories evolve by a kind of natural selection. Those that fail to predict the world are eliminated (or, more typically, revised) and those theories that stay with us are the ones that work so far. For example, the geocentric astronomy of Copernicus was accepted at one time but eventually replaced by the heliocentric theory despite religious resistance. There are simply too many facts that are consistent with the view that the Earth revolves around the sun and not the other way around: a good example of falsificationism in action.

The problem of axiomatisation

If science is really a process of logical argument, then psychology is in all kinds of trouble. For a theory to be treated logically, it must first be *axiomatised*. That is, it must be rendered as a set of logical propositions (or mathematical equations), as is the case in traditional physics. Then there can be no ambiguity as to its logical structure, and formal rules of inference (independent of the thought processes of the theory's author) can be applied to test its consistency and to generate predictions. Unfortunately, the bulk of psychological theories are presented as natural language arguments. Such arguments fall well short of the requirements of an axiomatised theory. There may be hidden or implicit assumptions. Words used in one sentence may take a different meaning than when used in another and so on. This is one of the main reasons that physicists sometimes accuse psychology of being a 'pseudoscience'.

I remember on one occasion deriving what I thought to be a clear prediction from the theory of a rival author. I was pretty sure that my experiment would falsify the prediction, so I emailed the author concerned to ask if he agreed with my prediction before putting it to the test. He refused to commit himself in advance of seeing the results. I remember at the time wondering if we were doing pseudoscience. If a well-known, published theory requires that its predictions are confirmed by the opinion of the theory's author, then this is hardly science at all. Had the theory been axiomatised, it would have been a matter of logic (or mathematics, which is a form of logic) to deduce the prediction that interested me. I find this just as worrying today.

One form of axiomatisation that seems to be returning to fashion in psychology is mathematical modelling. During the 1960's and 70's, there was a growth of interest in mathematical learning theory and other mathematical models such as signal detection theory (Tanner & Swets, 1954). The idea in this approach is not simply to predict trends that can be subjected to statistical analysis but to fit models to the actual data. For example, a mathematical learning model might predict the mean and variance of the number of trials taken to learn a task. A difficulty with such models is that they required parameters to be estimated from the data, and each parameter removes a degree of freedom. There are often insufficient free data points to provide an adequate test of the models and sometimes models based on completely different psychological principles will provide an equally impressive fit to the data.

The growth in mathematical psychology was arrested and to some extent reversed by the emergence of cognitive psychology in the 1970's. Exciting though the new information processing metaphor was, this was in some ways regressive. Far from trying to account for the results of psychological experiments in more detail, the new theories were quite remote from the data. Early cognitive models were linear in nature and often presented as flowcharts. They were strikingly deterministic while the data of the experiments remained intractably probabilistic. The theories would predict that people do A in condition 1 and B in condition 2. The data – at best – would show that the probability of A was higher in 1 than 2 and so on. Such findings, supported by statistical analysis, were regarded as confirmatory and nobody seemed to notice the discrepancy. This problem is still largely with us, of course. However, there has been a marked increase in publication of mathematical models, certainly in cognitive psychology, over the past 10–15 years.

A more recent form of axiomatisation is computer modelling. There is a range of computer modelling techniques available to psychologists. Monte Carlo models are explicitly statistical and produce probabilistic predictions. Rule based cognitive models operate deterministically and have to have human error built in to them in rather artificial ways. Neural network models adjust mathematical weightings in connectionist systems, largely to simulate learning processes. The last are often referred to as implicit models. They might simulate human behaviour but they do not render an explicit theoretical account of it: just a long list of numbers.

Computational modelling is still the exception in most fields. When used, it may achieve the objectives of axiomatisation. In particular, you have to render your theoretical assumptions into an explicit and complete form in order to write a computer program. The computation of consequences should also avoid human error in reasoning or tricks of natural language argument. Translating a verbal theory into a computational one can have unexpected consequences. An author in my own field who likes to do this is Phil Johnson-Laird, who implements his mental models accounts of reasoning in computational models written in LISP. He claims to have discovered a new set of phenomena called illusory inferences by this method (Johnson-Laird & Savary, 1999). He fed the program the premises of an argument and it generated a conclusion that he thought must be wrong. He spent hours checking for a bug, only to discover that the program had generated a valid but totally counterintuitive conclusion. Experimental tests then showed that subjects made illusory inferences on such problems.

In a rare collaboration with Johnson-Laird (Evans, Handley, Harper, & Johnson-Laird, 1999), we discovered something very interesting about one of his programs. We ran experiments on syllogistic reasoning and discovered a new phenomenon, the details of which are not important here. We asked Johnson-Laird to run the syllogisms through his program and – as we had suspected – the program was able to provide an account of the new finding we had discovered. However, we had difficulty formulating principles to account for this and none were apparently presented in the *verbal* theory that Johnson-Laird had published. What fascinated me about this was that the program had some explanatory power that the verbal theory it was supposed to be implementing did not. The two theories were not, in fact, equivalent.

Natural language arguments do have the major advantage that people can understand them! I have noticed that psychology is unusually verbose compared with most other sciences. I once published in a medical journal and was amazed by the brevity required. Such articles essentially take the form, 'this is what we did; this is what we found; goodbye'. The assumption seemed to be that science was a factual business and requires little interpretation. Nothing could be further from the way in which most academic psychology is conducted. The ability to publish research in top journals depends critically upon the skill of extended verbal argument in the introduction and discussion sections (see Chapter 7).

The problem of probability and the Bayesian alternative

Lack of axiomatisation is not the only reason why it is difficult to apply logicist (e.g. Popperian) philosophy of science to psychology. Propositional logical arguments apply to statements that are true or false. It is a deterministic system that does not allow for degrees of uncertainty. A theory is true or false. It either does or does not yield a specific prediction. The prediction is either corroborated or disconfirmed (absolutely). All of this provides a poor fit to psychology because ours is a statistical science. And yet many psychologists still feel that they are following a Popperian approach.

As noted above, few psychological theories have an explicitly statistical structure since the decline of stochastic mathematical modelling and yet we frame our predictions in statistical terms. We predict propensities of behaviour. We say, problem A is harder than B, so people will *tend* to take longer to solve it, or to make more errors. Or we say, people will *tend* to change their measured attitude more when the communicator has a middle class appearance and accent. And so on. We test such predictions by experiments that result in statistical significance testing (about which I will have much more to say in Chapter 9). If behaviour is measurably different in condition A than B, in the predicted direction, we regard our prediction as corroborated. If it fails, we regard it as falsified – maybe. We make such inferences with degrees of risk and uncertainty.

A technical weakness in the Popperian approach, apart from its deterministic character, is that it assumes the *evidence* is certain (see Howson & Urbach, 2006, for an extended critique of Popper). If the prediction of a theory is falsified, it does not necessarily mean that the theory is wrong. The *experiment* may be wrong in some way. Perhaps the conditions of observation were flawed; maybe a mistake was made in transcribing or analysing the data; perhaps the authors faked the findings. These things may be unlikely, but they certainly have non-zero probabilities. So if you can never be certain that the evidence is correct, you can equally never be sure that the hypothesis has been falsified. The use of statistical predictions magnifies this uncertainty. Type 1 and type 2 errors, for example, may occur. Perhaps the prediction failed because there was insufficient statistical power to detect the effect.

An alternative philosophy of science that is much better able to deal with uncertainty is that based on Bayesian principles (Howson & Urbach, 2006). Without going into the technical details at this point, I will try to convey the general approach. First, Bayesians take a subjective view of probability. A probability is no more or less than a degree of belief. If I am certain of an event, I assign it the probability 1; if I believe it to be impossible, I assign 0; if I am totally indifferent, I assign 0.5 and so on. Thus, it makes perfect sense to say that a theory or hypothesis has a probability. You believe it to some extent. Popper, who subscribed to the classical frequentist theory of probability, could not assign probabilities to theories or hypotheses – hence, the (unrealistic) obligation to accept or reject theories completely. Bayesians also envisage a process in which beliefs (probabilities) can gradually be revised in the light of evidence. The famous Bayes' theorem can be expressed verbally as:

Posterior belief = prior belief × diagnostic evidence

Prior belief is your current degree of belief in some hypothesis. When you encounter relevant evidence, this is diagnostic in some way. That is, it discriminates between the hypothesis and the alternative to it. Evidence can be strongly or weakly diagnostic and it can favour the hypothesis or go against it. Posterior belief is the probability you assign to some hypothesis after you have gathered the evidence. The theorem provides a multiplicative means of combining prior belief and diagnostic evidence. For example, if you have a very strong prior belief in the hypothesis and encounter evidence that is against it, but only weakly diagnostic, you will revise your belief downwards, but not very much. If you are indifferent to the hypothesis and encounter strongly diagnostic evidence, then your posterior belief will be mostly based on the evidence you have seen.

Few psychologists apply Bayesian statistics to their data, and fewer yet have heard of Bayesian philosophy of science. Yet, I contend, most psychologists are implicit Bayesians. They certainly act much more like Bayesians than Popperians, in my experience. Suppose authors are strongly committed to a theoretical position. What happens when they encounter an inconsistent result? Do they abandon their theory as good Popperians? I think we all know the answer to that. Do they perhaps lose a little confidence in the theory, a process that will continue if a series of such findings emerge? This seems much more likely.

Uncertain deduction

Together with my collaborator, David Over – a philosopher who has worked extensively on the psychology of reasoning – I have recently discussed the uncertain nature of real life deductive arguments (Evans & Over, 2013). The psychology of reasoning has shifted into a new paradigm in which the importance of traditional logic has been de-emphasised. Indeed, some authors have proposed that Bayesian theory should become the standard bearer for rational reasoning instead (Oaksford & Chater, 2007). In the new paradigm, we no longer ask people to judge whether conclusions to arguments are necessarily *true*, but rather to assign degrees of belief or probability. Nor do we ask them to assume that premises are true. As we suggested in this paper, the future lies in asking people to assign degrees of belief to the premises from which they reason.

These developments have led to some confusion as to whether psychologists are still studying deduction, as opposed to induction. If the conclusion is uncertain, then is the argument inductive? As we point out, this is not necessarily the case: arguments can be deductive but uncertain. Many of the arguments of the famous detective Sherlock Holmes as described by Conan Doyle are uncertain deductions. Holmes' arguments generally take a deductive form: his conclusions are true if his premises are true. But readers may nevertheless feel uncomfortable with the fact that he always turns out to be right. The reason is that the assumptions on which he based his conclusions were often uncertain in themselves.

While forming scientific theories involves inductive and abductive inferences, the theories themselves involve uncertain deductions. Predictions are usually derived by valid argument but the assumptions on which they rest are often conjectures that may or may not be true. Hence, when a prediction fails – and we are confident that the empirical evidence was sound – we may end up revising our belief in one or more premises, rather than discarding the whole theory. In fact, this is typically what happens. We revise aspects of the theory rather than throwing it out at one Popperian swoop and starting all over again.

Cognitive biases in reasoning and hypothesis testing

I assume that readers of this book want to be 'good' scientists, who are free from bias. When I said in the Introduction that good science and good careers tend to go together, I was referring to an imperfect correlation. There are counterexamples. In particular, the way in which researchers react to falsifying results may illustrate some divergence between scientific and career goals. Theories are important career (and self-image) builders and to concede too much too quickly when bad results are encountered might be seen as resulting in loss of face and prestige. The psychological conditions are certainly in place to allow bias to occur.

The cognitive and social cognition literatures have amassed a fair amount of evidence for two forms of bias that are relevant here. I define them as follows:

Confirmation bias: seeking to test a hypothesis in such a way as to discover confirming evidence and to avoid disconfirming evidence.

Belief bias: biased evaluation of evidence so as to preserve prior belief.

These two biases are clearly closely related in that they are both belief preserving. However, they are different. Confirmation bias (Klayman, 1995) is indicated by someone actively looking for supporting evidence and/or avoiding disconfirming evidence. Evidence for this can be found in social psychology. For example, people tend to seek out the company of others likely to share their views and beliefs and, correspondingly, to avoid those likely to disagree. People who have made a difficult decision, say to buy car X and not Y, may read advertisements for X's – and avoid reading ads for Y's – *after* the decision was made, in order to bolster belief in their decision.

Confirmation bias

Researchers are social beings to whom the laws of social psychology apply. I have already talked about the benefits of working in research groups (Chapter 6). However, this practice might also encourage confirmation and belief biases, or 'groupthink' (Janis, 1982), in which individuals seek consensus in order to maintain social cohesiveness of the group. It is certainly true that groups of researchers who work together over long periods of time tend to develop much common belief about the phenomena they study and end up in very much the same theoretical camp. When a difficult, potentially falsifying result appears, you will be likely to discuss it first with these like-minded colleagues, seeking reassurance and support. The major issue, however, is whether confirmation bias is likely to influence your choice of research design. If you accept the Popperian philosophy, then you ought apparently to be trying to find just the experiment or study most likely to *disconfirm* your theory. This has almost zero psychological reality in my experience. On the contrary, most researchers try hard to find an experiment that will support their hypothesis. The question is: does this matter? The answer is more difficult than you might think. Wason (1968) strongly believed that people were bad Popperians who were afflicted with confirmation bias, or 'verification bias' as he often preferred to say. While Wason was convinced that his experiments demonstrated such a bias, later researchers have cast considerable doubt upon this (see Evans, 2014). It seems much more likely that what he really showed was a *positive testing* bias.

Consider Wason's famous 2 4 6 problem. You are told that the experimenter has in mind a rule that applies to triple of three whole numbers. An example of a triple that conforms to the rule is 2 4 6. Your task is to discover the rule by generating your own examples. A typical protocol might look like this:

6 8 10	yes
10 20 30	yes
1 2 3	yes
100 200 300	yes

This participant has generated a series of triples, all of which conform with the experimenter's rule. She quickly becomes convinced that she knows the rule and at this point may say:

The rule is any ascending sequence with equal intervals

The experimenter replies that this is wrong and that she should continue testing triples. The participant continues testing as follows:

7 10 13	yes
12 10 8	no
111 222 333	yes
98 100 102	yes

At this point, she might do something very strange: announce a rule logically equivalent to the previous attempt such as:

The rule is adding a number, always the same, to make the next number in the series

Participants get incredibly frustrated on this task and baffled by the rejection of their repeated and rephrased hypothesis. The actual rule is 'any ascending sequence'. What has happened is that they have adopted a hypothesis that is too specific, so that any triple that conforms to their hypothesis also conforms to the rule. Hence, they can never refute their hypothesis by a *positive* test. A negative test such as '1 3 8, no' is required, but is incredibly difficult to find. It is far from obvious that positive testing bias is confirmation bias, however, as I defined it above. The participant does not know that positive tests can only confirm; that is a trick of this particular task. Thus, we do not know that they are *trying* to confirm their hypothesis, even though they are endlessly succeeding.

In real science, positive testing can usually produce disconfirming as well as confirming results and there are good arguments for supposing that a positive test strategy will usually be effective (Klayman & Ha, 1987). We think, 'if my theory is right, what can I predict should happen?' We then do an experiment to test if it does. We do not generally think, 'if my theory is right, what should not happen?' However, negative tests do arise sometimes. For example, we might consider what the *necessary* conditions for a phenomenon are. We might then conclude that removing such a condition would cause the effect to disappear. This has happened in my own work. For example, I came to the view that a phenomenon known as 'matching bias' was due to difficulty in understanding implicit negation. I figured that if I reframed the problems using explicit negations, the effect should disappear. I ran experiments that successfully supported this (negative) prediction (Evans, Clibbens, & Rood, 1996). Note, however, that this was a negative test intended to *confirm* my theory. Had the effect not disappeared, or been substantially weakened, the finding would have been disconfirmatory. Another difficulty with negative testing is our reliance on significance testing. While inferences drawn from significant findings are dubious enough (see Chapter 9), it is even harder to draw sensible inferences from a *non*-significant finding.

Do real scientists show confirmation bias? Research by Fugelsang, Stein, Green and Dunbar (2004) provides real world evidence showing how scientists behave when confronted with disconfirming evidence. Their study involved naturalistic observation of real research groups working in the area of molecular biology. Of 417 experimental results studied, over half (!), 223, produced findings inconsistent with the scientists' prediction. Analysis of videotaped research group meetings focussed on these inconsistent or disconfirming results. Did the scientists follow a Popperian approach by immediately giving up their hypotheses? In fact, they did this on only 31% of occasions. In the majority of cases, 69%, they blamed some aspect of the experimentation for the predictive failure.

According to Wason, these scientists are showing a confirmation bias. However, as pointed out above, a weakness in the Popperian system is the assumption that the experimental evidence is certain, when of course it is not. Whether you think the scientists' behaviour to be rational or not should depend on what happens next. Fugelsang et al. looked at what the scientists did in response to these inconsistent findings. Of the 223, 154 were followed up in further experiments that often modified or improved the methodology in some way. Of these, 84, or 54%, replicated the inconsistent finding. This shows that in many cases, the scientists were justified in their decision not to abandon the hypothesis. Moreover, when the inconsistent result was replicated, they were now willing in the majority of such cases (58%) to adjust their theoretical beliefs about the phenomenon. This process certainly looks much more like a Bayesian than a Popperian strategy. Bearing in mind that there might be much good prior evidence on which to base belief in the theories under test, the actual behaviour of the scientists is indeed quite rational from a Bayesian perspective.

From my own experience, I can easily imagine what is happening in these research groups. A hypothesis is established in a set of experiments or else follows from a well-established theory. An experiment is run in which the hypothesis fails. A good deal less cognitive work is required at this point to find a fault with the experiment than to revise the belief system that motivated the hypothesis. The first resort will therefore be to repeat the experiment with an improved methodology. If the finding persists, however, serious doubts start to creep in. At this point, you do not generally abandon the theory – which you may have spent years developing – but rather make *the least possible change* to it necessary to accommodate the inconsistent finding. As Kuhn (1970) observed, paradigms are usually brought down by criticisms from those outside rather than within.

Another important characteristic of how people test hypotheses is that they generally consider only one hypothesis at a time. This has been shown on a range of cognitive tasks, including classic studies of concept formation (Bruner, Goodnow, & Austin, 1956), studies of the 2 4 6 and closely related problems and work on so-called 'pseudodiagnostic' reasoning (Doherty, Chadwick, Garavan, Barr, & Mynatt, 1996). What seems generally to happen is that we form a hypothesis and subject it to test (generally using positive predictions) until or unless we encounter disconfirming evidence. In laboratory tasks, at least, participants generally give up hypotheses when disconfirming evidence is encountered. They then try to generate an alternative hypothesis that is consistent with previous evidence, so far as they can remember it.

There are philosophers of science who recommend simultaneous testing of multiple hypotheses but this seems to occur quite infrequently in real research. I think the reason probably has more to do with cognitive constraints than confirmation bias. When we engage in hypothetical thinking, we do so by use of an explicit thinking system that occupies central working memory and is of quite limited processing capacity (Evans, 2007a). Building a mental model to represent a single hypothesis is complex enough. What seems to happen is that we focus on the hypothesis that seems the most probable or relevant and then maintain it until there is good reason to give it up. In the jargon of decision theory, we *satisfice* rather than optimise. That is to say, we accept the hypothesis (choice, design) we are considering if it seems good enough. Optimisation requires systematic comparison of all logical possibilities: a practical impossibility in the real world.

Note that the preferred hypothesis may be the one that seems the most *relevant*. Relevance is subjective, as it relates the goals that individuals pursue and their personal set of beliefs about the world. It is commonplace in research to see that one researcher, or research group, favours hypothesis A while another favours B. Where evidence is less than decisive, which is normally the case at the cutting edge of research in any discipline, these rival hypotheses may be maintained by their supporters for long periods of time, sometimes for many years, sometimes for entire working careers. Each group sees enough supportive evidence to maintain their hypothesis and insufficient reason to give it up. This is perfectly normal science.

Consider two models of courtroom procedure. Some countries have an investigating magistrate (judge) who examines evidence and questions witnesses in a supposedly dispassionate effort to discover the truth. Other systems, such as those used in the UK and the USA, employ an adversarial system in which prosecutors and defenders argue opposing cases in the hope that evidence is fully tested and the truth will emerge in the minds of the jurors. People inexperienced in the workings of science might assume that scientists act like investigating magistrates. However, the adversarial model provides a much more accurate analogy to what actually happens. The rival positions taken by different theoretical camps motivate the conduct of empirical studies that hopefully shed new light on the issues. The jurors are played by the scholars who study the scientific literature. This adversarial system provides some protection against confirmation bias. You may not want to find a way of falsifying your theory, but there is always some rival researcher out there who will be only too pleased to do it for you!

Belief bias

Belief bias concerns the interpretation of data that has been encountered. This could be the results of your own studies or of those reported in the literature. In the deductive reasoning literature, it is defined as tendency to accept as valid the conclusion or arguments based upon whether or not you believe the conclusion. People are repeatedly demonstrated to accept more invalid arguments whose conclusions are believable and to reject more valid arguments whose conclusions are unbelievable, despite clear instructions to assume the premises are true and evaluate the logical necessity of the conclusions (Evans, Barston, & Pollard, 1983; Klauer, Musch, & Naumer, 2000). However, there is reason to believe that the bias is much broader than that. Researchers in the social cognition tradition have shown that people will be biased quite generally in their evaluation of evidence (Nisbett & Ross, 1980).

The most important way that belief bias can operate in scientific practice is by biasing responses to studies whose conclusions either favour or go against an established belief that the reader of the study holds. Suppose you are attached to a particular theory from which a particular hypothesis is derived. You read two studies in the literature: Paper A finds evidence for the hypothesis while B finds evidence against it. There is no question about your emotional reaction to this: you like A and you find B worrying or even threatening. If you exhibit belief bias, then you will seek to refute or discredit the evidence of B in some way. You might challenge the methodology or the sampling procedure or the analysis or the interpretation of findings. In one nice psychological study (Lord, Ross, & Lepper, 1979), students were asked to assess research studies using one of two methodologies and arrive at one of two conclusions: either for or against the deterrent value of the death penalty. Participants held strong prior beliefs on this issue. Despite full counter-balancing of the methods presented to different participants, they were far more critical of the methodology of the study reporting a finding contrary to their own beliefs than of one reporting congenial conclusions.

In Chapter 7, I discussed journal publication, including the behaviour of academic referees. I should point out here, however, that belief bias is definitely a factor with some referees. If your paper argues against their own preferred theory, you will usually encounter much fiercer criticism of your methods. Sometimes, powerful referees – that is, famous researchers – will succeed in suppressing for a while findings that run counter to their theories by hostile reports. However, journal editors should be wise enough and strong enough to know what is going on. Biased evaluation of studies is, of course, a principal cause of belief perseverance in the scientific literature.

Other relevant biases

Some other cognitive biases are worth mentioning briefly. There is much evidence that people are chronically overconfident in their judgements: that is, people overestimate how much they know (Griffin & Tversky, 1992). This bias applies to expert groups. Experts know more, but they still overestimate how much they know. For example, if an expert attaches a confidence of 80% to a prediction, she may only be 65% or so likely to be right. This could also explain why researchers often hold to their theories in the light of ambiguous evidence. Another important factor is hindsight bias or the knew-it-all-along-effect (Fischhoff, 1982). We chronically overestimate in hindsight what we could have predicted in advance. A common irritating experience for researchers presenting their findings is to encounter members of the audience saying, 'You didn't need to run that study – I could have told you that would happen'. Sometimes, it is a good idea to describe the design of your study and then ask the audience to predict the findings before you reveal them!

Overconfidence and hindsight biases are examples of metacognitive biases. Metacognition is what we know about our own cognition. The psychological literature is awash with evidence of failures of metacognition. For example, the reasons that people give for their actions in verbal reports are frequently not the causes that can be identified by behavioural research and may instead be after the fact rationalisations or theories applied to the self (Evans, 1989; Nisbett & Wilson, 1977). When people make judgements, the implicit policies that may be recovered by statistical means may diverge widely from the policies that the judge reports to the investigator (Doherty & Kurz, 1996; Harries, Evans, & Dennis, 2000) and so on. One consequence of these failures of metacognition is that experts – including scientists – may be systematically biased without being aware of the fact.

Finally, I should mention the 'planning fallacy' (see Bueler et al., 2002), an extraordinarily pervasive and persistent problem in all fields of life. People chronically underestimate the time that it will take to complete a task they are yet to start. There is some disagreement about the cause of the planning fallacy, but a plausible explanation is that people conduct a mental simulation of the task that is idealised and fails to envisage the practical problems that will affect most projects. What is really odd about this is that people seem not to make use of their experience of similar projects in the past. For example, experienced academic researchers may persist in making very optimistic estimates of the time required to write a research paper that does not correspond with the average time taken in the past. The planning fallacy is a potential problem in particular when preparing grant applications. If you specify more work than realistically can be achieved, then you may have problems explaining why you could not complete it in your end of grant report.

Conclusions

In traditional philosophy of science, logic and deductive reasoning are considered central to scientific thinking. Theories can be seen as logical systems allowing deduction of predictions and explanations. The popular (with psychologists) Popperian philosophy advocates falsificationism, which is based upon valid logical argument. However, I have shown in this chapter that the reality of psychological science is considerably removed from these philosophical discussions. Our theories are rarely axiomatised in psychology and hence we rely largely on natural language arguments rather than deductive logic or mathematics in order to predict and explain. The pitfalls of this approach have been indicated and we considered the advantages of one form of formalisation of theories in terms of computational modelling. However, this can also create its own problems.

In this chapter, I have also provided some discussion of the psychological nature of hypothetical thinking and several well documented cognitive biases that are associated with it. Of particular interest are confirmation bias and belief bias, as these tend to operate in such a way as to maintain belief in favoured theories and hypotheses. We may design the studies that will support rather than refute our theories and we may be quite biased in the way in which we assess studies published in the literature, since these biases seem highly pervasive in all human beings. However, science is a public activity in which different research groups and theoretical camps pit their wits in an adversarial battle. The history of science suggests that the truth usually comes out in the end.

I have also pointed out in this chapter the statistical nature of psychological science, which is one of the reasons why it can be hard to apply logical systems such as Popper's falsificationism. However, it does mean that it is important that we should understand the limitations of statistical thinking and judgement, which is another topic that has been extensively studied in cognitive psychology. The final chapter focusses in on this subject.

9 Statistical inference

All sciences have to deal with uncertainty in hypotheses and theories. If you subscribe to the Bayesian or subjectivist theory of probability, you can talk of these in terms of probabilities. In a statistical science like psychology, you have the additional problem of sampling error. Everything we measure is noisy and laden with error. If I want to measure your simple reaction time or your IQ or how long you take to solve an eight letter anagram, I will likely get a different number every time I try. I might therefore average a number of observations to estimate the 'true score' that underlies all the observed scores you are capable of generating. Of course, true scores vary between individuals as well. If I want to know the mean IQ of psychology students, for example, I have to worry about whether the sample of people I choose to test is large enough, or representative enough to give me an accurate answer. This is why psychologists have developed their toolbox of inferential statistical techniques.

How do psychological researchers cope with reasoning about all these uncertainties, aided as they are by their formal statistical training? Not very well, is the short answer, although to be fair, no worse than others working in statistically based sciences. Let us start by clarifying what we mean by probability. There are several philosophically distinct theories of probability of which the most important are the frequentist and the subjectivist (or Bayesian) theories. Conventional statistical training tends to convey the 'scientific' definition of probability as long-term frequency. For example, to say that a coin has a probability of heads of 0.5 means that in a long sequence of tosses, it would land on heads about 50% of the time. This view has at least the superficial attraction of being 'objective'. Using this theory, we teach our students about sampling distributions. We say, imagine that we were to repeat a given experiment, with the null hypothesis true, a large number of times – how often would this particular result arise by chance? This probability appears to be the relative frequency with which an event occurs.

There are technical objections to the frequency theory. In particular, an observed proportion tends towards the theoretical probability in a long run but is never asymptotically equal to it. Even if we tossed a perfectly fair coin an infinite number of times, the proportion of heads would still not have to be equal to 0.5. That is the beauty of probability – anything can happen. To my mind, a much more serious objection is that we can only talk of probabilities for events that have frequency distributions. This is a horrendous limitation. Consider all the one-off but risky decisions you make in a lifetime: whether to accept a job offer, whether to get married, whether to buy a particular holiday package. In general, we have no frequency data to draw on[1] and yet we still assess the risk involved in such decisions. Will I enjoy working in this environment? Will I still want to spend my life with my spouse to be in 10 years' time? Will the holiday resort prove to be noisy and crowded? Intuitively, these decisions feel risky because they involve uncertainty and the outcomes matter to us. Our choice ultimately depends on the degree of probability that we mentally assign to key events, even though these are unique and belong to no frequency distribution.

The Bayesian notion of probability, briefly introduced in the previous chapter, is subjective or psychological. A probability is neither more nor less than a degree of belief in something. I assign an event a probability of 1 if I am certain that it will happen, 0 if I am certain that it will not happen and intermediate values for intermediate degrees of confidence or belief. This subjectivist approach means that probabilities can be assigned to anything, including scientific theories and hypotheses. We can talk of the probability that the universe is infinite, or that there is life on Mars or that robots will do all the housework in the year 2112. Those who advocate the frequentist or 'classical' approach to probability object that the subjectivism of the Bayesian approach makes it unscientific. Hypotheses can have probabilities, but these can be different for different scientists.

Let us return to an issue introduced in the previous chapter: the problem of induction. Recall that the difficulty was that we had no logical means of inferring a general scientific law from any set of particular observations. To understand the statistical version of this problem, let us think about the evidence (or datum D) and the law (or hypothesis H). The statistical problem of induction is that we would like to know how probable a hypothesis is given the evidence: formally, we want $P(H|D)$. Unfortunately, there is no really satisfactory way to determine this. Suppose I tell you that a coin was tossed 10 times, resulting in the following sequence:

HHHHHHHHHH

That is right – it came up heads all 10 times! Now, I ask you: what is the probability that the coin used was fair – that is, it had an equal probability of coming down heads or tails on each toss. You might think that you can demonstrate beyond reasonable doubt that the coin is not fair. You say something like this: the chances of a fair coin landing heads 10 times in a row is 2^{-10}, or approximately a probability of 1 in 10,000. This probability is very low, so we can say beyond reasonable doubt that it was not a fair coin. Unfortunately, you have not answered the question that I asked. What you have calculated is $P(D|H)$ – the probability of the observed data given a fair coin. What I asked for, however, was $P(H|D)$ – the chance of the coin being fair given the observed sequence. The confusion of these two probabilities I call the *conditional probability fallacy*. It is a fallacy that underlies many errors of statistical thinking, as I shall demonstrate.

With wounded pride, perhaps, you try a riposte. 'All I did was to test a null hypotheses and reject it, just as I was taught in stats classes. If this was a fallacy, then statistical significance testing must be a fallacy too'. 'Exactly', I reply . . .

Statistical significance testing

Statistical significance testing, as preached in the textbooks on psychological research methods, is – in my opinion – a complete nonsense. This is something that has been known to many people for many years. So why do we still practise it? As Gigerenzer and Murray (1987) – who explain the problem as well as anyone – put it:

> It is remarkable that after two decades of . . . attacks, the mystifying doctrine of null hypothesis testing is still today the Bible from which our future research generation is taught.

To understand the problem, let us consider a statistical fallacy that arises in a quite different context. In his book on risk, Gigerenzer (2002) discusses a phenomenon known as the 'prosecutor's fallacy' (Thompson & Schumann, 1987). This is an erroneous statistical argument, beloved of prosecuting lawyers – and their expert witnesses – in criminal court cases. To illustrate the fallacy, consider the following scenario. A woman's body is found dumped in a car park in a large city: she has apparently been raped and murdered. Semen is found inside the body, yielding a DNA profile. After failing to trace a suspect by conventional methods, the police initiate widespread DNA testing of men in the surrounding area. Eventually, a man is found to have a DNA match and, unable to account for his whereabouts at the time of the murder, is arrested. In spite of having no other evidence to link the suspect to the crime, he is charged with murder and put on trial.

At the trial, the prosecuting counsel calls an expert witness who testifies that the chance of the DNA match occurring by chance is less than 1 in 100,000. The prosecution argue that this probability is so low that it proves beyond any reasonable doubt that the defendant is guilty as charged. Sounds convincing, right? However, even if we leave aside other relevant probabilities, such as that of laboratory error in the DNA test, or of the suspect having had consensual sex with the woman who was later murdered by someone else, it is a fallacy. The random match probability is actually a conditional probability. Specifically, it represents:

P(DNA match|innocent suspect)

The fallacy is to equate this probability with the probability of innocence, arguing in effect that the defendant is 99.99% likely to be guilty. When we do this, we commit the conditional probability fallacy: confusing the $P(A|B)$ with its converse, $P(B|A)$. In this example, the relevant probability is actually:

P(innocent suspect|DNA match)

In Bayesian terms, this is a posterior probability: that is, the probability of a hypothesis after some evidence has been inspected and this depends upon the prior probability or base rate. Suppose a reasonable estimate is that any of one million men might have committed the crime described. Of these, we could expect about 10 to show a match (1 in 100,000). Hence, in this case, the probability of someone with a match – and *no other evidence* to link them to the crime – actually being innocent is 9 out of 10 (see Figure 9.1).

In case this makes you worried about innocent people being convicted on DNA evidence, it actually gets worse. This is the probability that would apply if just one individual is taken at random and tested. If we take into account the police practice of testing many men at random, the probability alters substantially in favour of a positive match with an innocent man. Suppose 10,000 *innocent* men are tested, each with a random match probability of 1 in 100,000. The chance that one or more of these will show a match comes down very close to 1%. This is *before* we take into account the base rate.

Does the structure of the prosecutor's argument seem somehow familiar? It should. We (academic and research psychologists) read it in our journals every day and teach it to our students every week. The way that psychologists are taught to interpret statistical significance levels embodies precisely the same fallacy. It goes something like this. Construct an experimental hypothesis (H_1) and a null hypothesis that is its complement (H_0). For example, H_1: there is a difference in the mean performance level of two experimental groups; H_0: there is no difference. Compute the probability of the experimental findings under the null hypothesis. If this is below an arbitrary probability value, known as the significance level, then reject H_0 and *accept* H_1. The fallacy lies in the last bit. From a significance level alone, we can make no inference at all about the probability of H_1. Again,

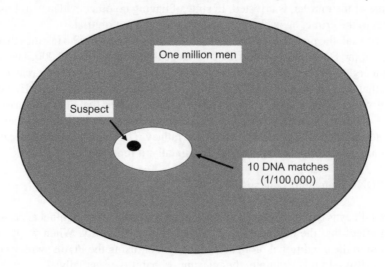

Figure 9.1 The prosecutor's fallacy

the problem lies in conditional probabilities. The significance test yields the following probability:

P(data|null hypothesis)

Whereas the probability we are interested in is the posterior probability:

P(null hypothesis|data)

For null hypothesis, read innocence and for data, read DNA match. It is exactly the same argument. As already mentioned, psychologists' fallacious use of statistical significance testing has been known for many years, although the profession has resolutely ignored attempts to remedy this. As Gigerenzer and Murray (1987, p. 25) put it:

> Psychologists do not wish to be cured of these illusions, for the consequence would be the abandonment of the indispensable instrument and hence the abandonment of the unification of psychological methodology.

If anyone is still in doubt, engage in the following thought experiment. Suppose a large bag contains 1,000 coins: 999 of these are known to be perfectly fair, but one is known to be heavily biased towards heads. One coin is drawn out of the bag at random and tested by being tossed 100 times, with the number of heads and tails recorded. This results in a preponderance of heads. A statistical significance test is applied, using the binomial test, with the result significant at 1%. What is the chance that the coin selected was the biased one? According to the prosecutor's fallacy, the answer is 99%, but it should be pretty obvious that this is not right because there was only a 1 in 1,000 chance of drawing the biased coin for testing. We can find the correct answer by application of Bayes' theorem. One form in which the theorem can be written is:

Posterior odds = Prior odds × likelihood ratio

The prior odds reflect the probability of the hypothesis before any evidence is examined. Since there was only one biased coin in a bag containing 1,000 coins and since the coin was selected at random, these must be 999 to 1 in favour of the coin being fair. The 'likelihood ratio' refers to the extent to which the evidence discriminates among the hypotheses. Specifically, it is the probability of the evidence given one hypothesis divided by the probability of the evidence given the other. The significance test shows us that the probability of the evidence (sequence of coin tosses) given a fair coin is 1 in 100. So the likelihood ratio is 99 to 1 in favour of a biased coin. When we multiply these two ratios together, we find that the odds are 111 to 11 in favour of the coin being fair, despite the significant result found against it. The probability is 11/122 = .09.

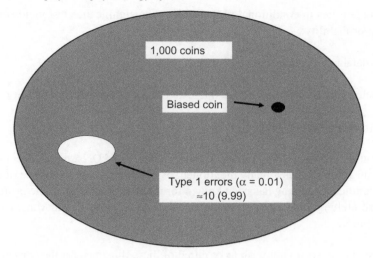

Figure 9.2 Significance test on one coin from the bag of 1,000

A more intuitive way of deriving the probability may be made by inspection of Figure 9.2. Here, we can see that if you tested 1,000 (strictly, 999 because one is biased) fair coins at the 1% significance level, you would come up with about 10 (9.9) false positives, or Type 1 errors. So there is a 1/1,000 chance that you picked the biased coin out of the bag and a 9.99/1,000 chance that you picked a fair one. Hence, the probability that your coin is biased is 1/10.99 = .09. Translating this into real research practice, what it means is that if you test for an effect that is unlikely to be present, then it is much more likely that a significant result is, in fact, a Type 1 error. This is one of the reasons why people who are sceptical about ESP are often unimpressed by published reports of significant findings. For example, they wonder how many *non*-significant studies were run and never published. It also shows the importance of replicating novel and surprising results *before* they are published.

Evidence of the conditional probability fallacy abounds in the cognitive psychological literature. For example, Casscells, Schoenberger and Graboys (1978) gave physicians the following problem, asking them to estimate the positive predictive probability:

> If a test to detect a disease whose prevalence is 1/1000 has a false positive rate of 5%, what is the chance that a person found to have a positive result actually has the disease, assuming that you know nothing about the person's symptoms or signs?

Few physicians gave the correct answer to this question and a typical response was 95%. This answer again confuses two conditional probabilities. There is a 5% chance that someone will have a positive test result if they do not have the disease. The doctors seem to think this implies its converse: that the (posterior) probability of not having the disease given a positive test result is 5%, in which case their answer that 95% of those with a positive test result would have the disease would actually follow. In fact, the correct answer to the question set is less than 2%. Of 1,000 people, one will have the disease and test positive: a true positive. Of the remaining 999 healthy people, approximately 50 (5%) will test positive: the false positives. So, of every 51 people with a positive test result, only one will have the disease. Mass screening for infrequent conditions (for example, mammography of women over 50) will almost always yield false results in the majority of those who test positive.

So what exactly does the statistical significance level mean? At best, it represents the degree of confidence that you can have in rejecting the null hypothesis (relative to what, ask the Bayesians?). What should we do if our test is non-significant? Do we infer that H_0 is true and that H_1 is therefore false? No, we teach our students – that is a fallacy. Just because we fail to reject the null hypothesis does not mean that it is true. We may not have had enough power to detect the effect, for example. Journal editors are reluctant to accept papers based on non-significant findings because they feel that these are inconclusive (unlike, apparently, significant ones!). If the object of significance testing was simply to reject null hypotheses, however, why bother? They are all false anyway! A null hypothesis is a point hypothesis, specifying a particular point in a continuous measurement system – for example, that the difference in mean scores between two groups will be exactly 0.00000 extended. No such hypothesis can be true. Whether you find a significant difference will, however, depend upon both how big the deviation from the null hypothesis really is (effect size) and the degree of sensitivity or power in your statistical procedure.

Power and effect size

Bayesian statistical analysis can be (but rarely is) applied in psychological research. To use it, you must first specify two alternative hypotheses (real alternatives – not a null hypothesis and its complement). You can determine posterior odds, but only by assigning prior probabilities to the hypotheses, which is often seen as a subjective process. What psychologists *are* increasingly doing is performing power calculations that involve the first of these two steps: specifying an alternative hypothesis. A most welcome development in editorial practice in recent years is the encouragement or requirement for authors to publish effect sizes and/or report power analyses. To do the latter (the Neyman-Pearson method), you have a null hypothesis H_0 as before, but now you *specify* an alternative H_1. For example, H_0 may be that there is no difference in the mean score between two groups and H_1 may be that there is a difference of one standard deviation. By this means, we can define *effect sizes* that are of theoretical or practical interest and set power levels appropriate to detect them.

When thinking about power and effect sizes, it is useful to note the relationship these have with significance levels:

(a) Observed significance is proportional to effect size, *with power held constant*
(b) Observed significance is proportional to power, *with effect size held constant*

One of many common fallacies of statistical thinking that psychologists commit is to overlook the qualifying condition of statement (a) above. That is, they assume that a large significance level – say, $p < 0.0001$ – indicates a large effect size. However, it would only indicate this if power characteristics were moderate. Serious errors of inference may arise from this. For example, you may read reports of two separate studies testing a drug for its effectiveness in lowering blood pressure. Both studies report significant effects but Study A demonstrates the effect of drug X at a higher significance level than Study B finds for drug Y. You conclude that drug X is the more effective, whereas the reverse might actually be true if Study B were conducted on a much smaller sample.

Application of the Neyman-Pearson method means that, in a particular study, we may define H_1 as a medium size effect and calculate – without any fallacy – that *if* an effect this size or larger is present, then we have at least an 80% chance of finding a statistically significant result. Such power calculations are most useful when results are *not* significant because then we can say that the effect tested for was probably absent (or too weak) to be detected. It does not, however, help us to decide whether a significant result was, in fact, a Type 1 error or not. For that, we still need the prior probability of H_1. As I illustrate with an example below, the significance level is the *worst case* estimate of the Type 1 error level: that which applies when no measurable effect was present at all. The practice of reporting effect sizes is welcome because these involve no statistical fallacy and can also show if we have found a trivial effect because power was *too high*. An example of an effect size with which we are all familiar is the correlation coefficient: we are always interested in the size of this, not just whether it was statistically significant. If in a large sample we get significant correlation but with a coefficient of less than .2, for example, we will find it of little theoretical interest.

So the mathematics of the Neyman-Pearson method are sound as far as they go, but the inference problem remains. A full Bayesian analysis would still require specification of prior probabilities for the hypotheses and also assignment of utilities to represent costs and benefits of getting decisions right and wrong. Power probabilities, just like significance levels, are *conditional* probabilities. A Type 1 error (false claim of significance) can only be made when the null hypothesis is true and a Type 2 error (failure to detect an effect) can only be made when the alternative hypothesis is true. The potential for misunderstanding the conditional nature of the statistical significance probability is illustrated by the following question:

> Suppose an experimental psychologist conducts many experiments in a long and distinguished career. She always reports as statistically reliable findings supported by significance levels of 5% or better. Over her career, how many Type 1 errors will she make? In other words, in how many of these experiments will she declare a result significant when in fact the null hypothesis was true?

The correct answer is that you cannot tell without further information. However, many psychologists will give the answer 5% (Pollard & Richardson, 1987), thus confusing the conditional probability with an absolute probability. We can look upon the significance level as a relative false positive rate, but the question demands an absolute false positive rate. The question asked requires consideration of base rates – in this case, the prior probability that the null hypothesis is true. The answer could only be 5% if this psychologist was unfortunate enough never to have run an experiment where an effect was present! Suppose that, in fact, effects are present (the null hypothesis is false) in 80% of the experiments that she runs. Type 1 errors cannot occur in any of these experiments because they are conditionalised upon the null hypothesis being true, which it is not. They could only occur in the 20% of experiments in which there was, in fact, no effect present. With a significance level of 5%, we could expect one such experiment to produce a significant result. Thus, the absolute false positive rate for this psychologist would be 1%, not 5%. She will commit a Type 1 error, on average, once in every 100 experiments. The structure of this problem is illustrated in Figure 9.3.

We could instead ask for another relevant posterior probability – the absolute rate of failures to detect effects (the miss rate). How often when she finds a significant result was the null hypothesis actually true? The conditional probability fallacy would again yield the answer 5%. To answer this accurately, we need to know the Type 2 error rate (a *conditional* probability) or the number of times that she fails to detect a significant results *when* an effect is present. This depends on the statistical power employed, but let us say, for the sake of argument, that Type 2 errors occur in 25% of her experiments when an effect is present. So the answer to the question, given these estimates, would be as follows: out of every 100 experiments she runs, 80 have a true effect. In 75% of these cases, she has sufficient power to declare a significant result, leading to 60 true positive findings. In 20 cases where there is no effect present, she will declare a Type 1 error or false positive about once, since the significance level is 5%. There are hence 60 true

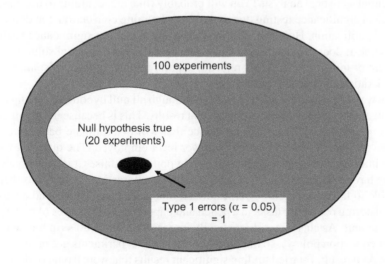

Figure 9.3 Number of Type 1 errors made by a career researcher

positives and 1 false positive. So the posterior probability of missing an effect is 1 in 61, or 0.016%, rather than the 5% that might typically be suggested.

Do psychologists have good intuitions about statistical power? There are some studies that provide direct evidence on this. Tversky and Kahneman (1971) carried out experiments on experimental psychologists and concluded that they held a belief in 'the law of small numbers'. This is the belief that the law of large numbers applies to small numbers as well! To put it another way, psychologists believe that small samples generate more information about populations than they actually do. In general, studies of intuitive statistical judgement show that while people have some sensitivity to sample size, they systematically underestimate its influence (Evans, 1989; Kahneman & Tversky, 1982). I once ran some experiments with Ian Dennis (Evans, 1992) in which we asked psychology undergraduates with a minimum of 12 months' statistical training directly to estimate the power characteristics of research designs given specified samples and effect sizes. Judgements were extremely conservative on both factors. That is, the students' judgements greatly underestimated the effect of both sample size and effect size on power.

A different approach is to look at what psychologists actually do in their published research. Cohen (1962) carried out a survey of psychological experiments published in the *Journal of Abnormal and Social Psychology* and claimed that the power characteristics of the published experiments were generally inadequate. Sedlmeier and Gigerenzer (1989) repeated the exercise many years later to see whether Cohen's paper had any effect on the design of later psychological experiments. Needless to say, it had none at all!

Practical implications

What are the implications of the above discussion for how you should conduct inferential statistical analysis? You will probably (like me) continue to use classical statistical significance testing due to the overwhelming convention that dominates journal publication. However, it is well to be aware of what significance testing is and is not. It does provide diagnostic evidence but is not a magical solution to the problem of induction. Nor does rejecting a point null hypothesis in isolation from consideration of alternatives make any sense.

The way I like to think about it is this. Although all null hypotheses are false, not all statistical testing will lead to significant results. This is because the significance test is in effect a decision making device that will discriminate between effects of a given size, depending upon the power level employed. The question we are really interested in is not whether the null hypothesis is false (it is) but whether or not we have evidence of an effect of a size that is interesting. This means that it is possible to have *too much* power. I noticed this in the past when I had a habit of administering reasoning tasks in a pencil and paper format to classes of undergraduate students. As class sizes dramatically increased in the 1980's (in line with UK higher education policy), so the sample sizes in my experiments got progressively larger. As a result, I started finding significant results that were trivial or downright irritating (for example, higher order interactions between factors). When I realised what was happening, I started splitting the groups and running two experiments at the same time! Of course, there are statistical techniques (particularly in multivariate designs) where large samples are needed.[2]

This is not a research methods text, let alone a statistics book, so I am not going to give advice on particular statistical techniques. You should, however, acquire the habit of measuring effect sizes or performing power calculations as it is good practice, one increasingly encouraged by journal editors, and helps you understand your data a lot better. However, prior to the past decade or so, the great majority of psychological experiments have been conducted without these precautions, reporting only classical significance testing. Does this mean that they are based on fallacious inference? Essentially, yes. Does it mean that the results are meaningless? No. I guess I had better explain that.

What I think actually happens in practice is this. Researchers in particular fields discover effects that are of theoretical or practical interest. Depending on the field, such effects are roughly of a particular size. The power characteristics are set by custom and practice and endlessly copied by other researchers. This is the most commonly used heuristic for research design (see Chapter 3). For example, researchers using psychometric methods use larger samples than those using experimental methods. Thus, by trial and error leading to convention, power is set not so low that it will miss effects of interest and not so high that it will lead to significant but trivial results. Hence, as a decision making tool, statistical significance testing in practice is useful for detecting effects of interest, even though its ostensible purpose of rejecting null hypotheses is nonsensical.

Those of you who have followed the arguments in this chapter carefully will be aware that the problem of prior probabilities remains. It is still invalid to infer that your alternative hypothesis is established due to a significant result, as you are equating the posterior probability of the alternative hypothesis given the data with one minus the likelihood of the data given the null hypothesis. From a Bayesian perspective, the posterior odds can only equal the likelihood ratio if the prior odds are balanced. That is, if you are indifferent, *a priori*, to whether or not the null hypothesis is true. Thus, a significant result might increase your degree of belief in a hypothesis and a non-significant one decrease it, but the posterior belief will also be related to the prior belief, just as it should be, according to the Bayesian philosophy of science discussed in the previous chapter.

It is far from satisfactory that we are muddling along with an ill-defined method of inferential statistics that invites fallacious reasoning, even if the conventions of research fields tend to make statistical significance a reasonable indicator that effects of interest are present. I think one thing that would help more than any other is to understand and think far more about the effect size and a lot less about statistical significance. When I get a new set of data, the first thing I do is to compute means or draw graphs so that I can get a picture of what is going on. Are the trends as expected? Do they look big or small? Are there other factors affecting the data that I did not expect? Significance testing is secondary. If effects are evidently small, then I am not very interested in them, whether they are significant or not.[3] If interesting effects seem to be present but not significant, then I worry about whether I have used the right power characteristics or the right method of analysis. It always makes a big difference whether the results make sense *theoretically*. I guess this is where the Bayesian priors come in.

I once had a PhD student who discovered a significant four-way interaction. This worried him for months. He would come to coffee and pull from his pocket his latest drawing (of dozens) to show the interaction in the hope of finding an interpretation. I told him to forget it, that if he repeated the experiment, it almost certainly would not be significant again and even if it was, it did not matter. This obsessive level of attachment to statistical significance is very common but, frankly, in the order of magical thinking. Even in its own dubious terms, significance testing is *risky*. Type 1 and Type 2 errors will occur. It is simply not the case that a result that just passes the threshold for significance is indisputably real, whereas another (slightly smaller) effect that just fails to reach significance is pure random error.

Thinking like this can lead to another fallacy that I have not mentioned yet: that of imaginary interactions. This is committed by many experienced researchers, not just by students. For example, you test for an effect in two groups: say, males and females. The result is (just) significant for the males and (just) non-significant for the females. Many people will discuss these results as though they show an interaction: the effect is 'present' for males and 'absent' for females. Of course, if you carry out an interaction test, it is not remotely significant because the difference in the differences is so small. Those who think magically about significance levels find such results paradoxical and confusing.

Tversky and Kahneman (1971) reported a related illusion. They showed professional psychologists two data sets with similar trends. When assessed separately, they each fell short of significance, but when combined, they were comfortably significant. One group were told that the second data set was an extension of the first study and judged the effect to be highly reliable. The other group were told that the second data set was from a replication study. Since both the original study and the replication were non-significant, they were confident that no effect was present. In other words, the two groups drew precisely the opposite inference from the same total set of data, depending upon how the problem was framed. The technique known as *meta-analysis* can be used to assess the overall evidence from a series of similar studies. This method combines analyses rather than examining whether individual studies are significant.

In summary, my advice is to make sure that you understand the relationship between power, effect size and significance, and that you think about all three when examining your data sets, rather than just focussing on statistical significance. Measures of effect sizes are now routinely available from the major statistics packages, such as SPSS. You should also look at methods that will allow you to measure power and explore techniques such as meta-analysis that give you a more reliable picture by combining data from a set of related studies. Do not think magically about significance: that is, do not assume that significant effects are real and present, and non-significant effects are really absent. Bear in mind that 'reliable' effects can really only be established in a series of studies whether they are repeatedly observed under a range of conditions. Do not be ashamed of your natural Bayesian tendency to revise your belief in hypotheses gradually as you encounter relevant evidence.

Cognitive biases in statistical judgement

Considerable research has been conducted on the biases and fallacies that affect people's thinking about probability. I have mentioned some of this work already, but will point the reader towards some other relevant literature before concluding this chapter. The most famous line of work is that of the 'heuristics and biases' programme, initiated by Amos Tversky and Daniel Kahneman (for collections of papers, see Kahneman et al., 1982; Gilovich et al., 2002). The psychological work already discussed above indicates clearly that psychologists and other professionals (e.g. medical researchers) are as prone to these biases as anyone else. I have provided detailed review and discussion of statistical thinking elsewhere (Evans, 2007a, Chapter 6) and will here just briefly mention some of the main biases that are most relevant to researchers in a statistical science like psychology.

Sample size neglect

As mentioned earlier, there is a chronic tendency for people to give too little weight to the size of samples, although they do not neglect this altogether (Kahneman & Tversky, 1982). This has been attributed to the idea that people judge the probability of samples by how *representative* they appear to be of the population (Kahneman & Tversky, 1972). Thus, people are most impressed by the similarity of the population and sample means, but cannot compare sample size, as this is not a feature of the population. However, when the size of the population is specified, as well as that of the sample (Bar-Hillel, 1979; Evans & Bradshaw, 1986), another fallacy is demonstrated. People believe that a sample of a given size is more valuable if the population is smaller, as if the proportion of the population sampled was relevant. It is not.[4]

The main relevance to psychological research here is to rely on formal calculations of power and not on intuition to determine your sample sizes. As discussed earlier, there may be conventions about sample sizes in your field that are set to capture findings of interest. But there is no harm in checking out formally whether these are appropriate.

Base rate neglect

Another classic finding of Kahneman and Tversky (1972) is that people tend to severely underweight base rate probabilities when making posterior probability judgements. This finding is sometimes mis-reported as a finding that base rates are ignored altogether (Koehler, 1996). In fact, a number of individuals do ignore base rates but a minority use them as the only basis for their judgements when individual differences are reported (Cosmides & Tooby, 1996; Evans, Handley, Perham, Over, & Thompson, 2000). The real problem is that people find it very hard to integrate information about diagnostic tests and base rates. This literature also shows how vulnerable expert groups can be to statistical fallacies, with doctors often failing to understand the importance of prior probabilities when interpreting diagnostic tests. It has also shown that presenting problems with frequencies rather than probabilities can make them easier to understand and solve, although there has been much controversy about the reasons for this (Barbey & Sloman, 2007).

Subjective randomness

In general, random sequences do not look random to ordinary people, and what they think is random is not. The bias is that we expect runs of particular outcomes to be shorter than they actually are (Evans, 2007a; Tversky & Kahneman, 1971; Wagenaar, 1988). One consequence is the 'gambler's fallacy': the belief, say, that if red comes up several times on a roulette wheel, then black is more likely to come next. This is technically known as *negative recency bias* because the individual predicts against the current run. However, failure to understand randomness can also lead to a *positive recency bias*. This happens when a person concludes that a sequence is not random but systematic. For example, if a run of reds leads you to believe that the roulette wheel is biased, you may predict red again. Gamblers actually have many false beliefs about chance and probability that help to maintain their losses in the casinos (Wagenaar, 1988).

In Chapter 4, I mentioned that you can have a run of successes or failures with grant applications and that this can have profound psychological effects. It is hard to accept that such runs are just the natural outcome of a chance process, but they probably are. We have a very strong tendency to see pattern where there is randomness and to attribute causality where this is only chance. Even when psychologists use a random number generator to order a sequence of problems in an experiment, they can be disconcerted by the apparently non-random nature of the outcomes. Some even fiddle the orders to make them look more random!

Availability and illusory correlation

Tversky and Kahneman (1973; for review of later work, see Reber, 2004) famously showed that people judge events to be more probable when they are more 'available' from a psychological point of view. This can arise because it is easier to generate examples in our minds, we are biased by selective media coverage, or we recall examples that were easier to remember because they were more vivid in some way (Nisbett & Ross, 1980). It could explain a bias known to affect professionals: *illusory correlation*. When we have a false belief that two variables A and B are correlated, we selectively remember cases where we observed that A and B go together. These confirming cases are hence more available to memory and bias our judgement (Chapman, 1967). Availability biases can affect scholarship. Papers that have striking findings, ones that fit our theories, use vivid coloured presentations, or are written by famous authors can all stick more strongly in our memories, creating the illusion that certain conclusions are more strongly supported by the evidence than they actually are.

Conclusions

Thinking about probability is difficult for everyone and we psychologists are not immune from the cognitive biases that abound when people make intuitive judgements and inferences that are statistical in nature. In fact, all the evidence suggests we are highly prone to them. You would think that our use of formal statistical procedures would provide adequate protection but, as I have shown here, the favoured statistical method of classical significance testing actually encourages a dangerous fallacy. It makes us think that we are finding evidence for hypotheses by simply rejecting a null hypothesis that is almost bound to be false in the first place.

It is not my purpose in this chapter to dissuade researchers from running significance tests, as they are entrenched in the practice of empirical psychology. I have been aware of the problems discussed here for many years, but have continued to report significance levels. It is my purpose, however, to make you more aware of what it is you are actually doing and to minimise the risk of false inferences. For many reasons that I have explained, significance testing can work quite well in practice even if it is flawed in principle. And the trend towards requiring report of effect sizes and power levels has certainly moved practices in the right direction.

Final thoughts

Before committing to a career in academia, it is well worth considering other options. There are many very smart people enjoying good careers outside of the academic world. Some of them make a lot of money and many will derive satisfaction from their work in industry or in government departments. In many ways, a university career is less attractive now than it was when I was starting out. Certainly in the UK, the age of austerity heralded by the economic collapse of 2008 has resulted in frozen salaries, poorer pensions (at higher cost) and greater workload for academics, along with other public sector workers. I imagine that similar pressures are being felt in different ways all around the world, as the economic issues are global.

Academia has changed over my career in other ways, not all of them good. Assessment and performance measurement are everywhere. Students seem much more interested in the marks and grades they obtain than in the quality of the educational experience. Academic staff (or faculty) are constantly assessed on their teaching performance, their research grant income and their publications. While research work ought to be – and can still be – *fun*, it comes with a lot of pressure to succeed in quantifiable ways. We are no longer seeking truth and enlightenment, it seems, but rather large research grants and papers in journals with high impact factors. To an extent, these pressures have increased the quality of research work; they have certainly increased its *quantity*. But something important seems to have been lost in the process.

Having said all this, there is still something about the life of an academic researcher that strongly attracts particular individuals. I have known colleagues in subjects like chemistry and geology to turn down offers from industry that would have *tripled* their salaries. Why? Because they want to own their own work and their own ideas. If you work in industry or in a government department, then you will not own the ideas – technically and legally, the intellectual property rights (IPR) – in your own work. Even if you do research, it will be to achieve the aims of your organization, not your own. You will sign confidentiality and non-disclosure agreements. You may be highly restricted in what you can publish. And if you leave, all of your intellectual work is left behind with your employers. Nothing can go with you except for the skill and expertise you have developed.

What is unique to the career of an academic researcher is the personal expression, intellectual freedom and complete ownership of your intellectual property. All the research you do can and should be published openly for all the world to read. If you move from one university to another, every idea you developed goes with you, together with your body of published work and your academic reputation. Your career is essentially independent of your current employers. I officially retired a few years ago, but it initially made little difference. Although I was no longer required to teach or attend meetings, I found that many of the demands on my time were generated externally and continued as normal. I continued to get many requests, for example, to write papers, give talks, review articles and so on. Of course, employment is required for the salary, the contact with colleagues and students and the access to research facilities such as laboratories. But I cannot think of another profession in which your personal career is to such an extent independent of your current appointment.

The current book is intended as both a philosophical and a practical guide to the business of doing academic research and its contents reflect mostly the hard-earned experience of one who has been doing all of these things for a very long time. You, the reader, may be a final undergraduate considering whether to apply for a PhD and pursue a research career. You may be a research student or postdoctoral researcher who has started down the road but is not yet finally committed to a career in academic research. If so, I hope that Part One of this book has helped you decide whether the life attracts you personally. If it has, you should also be aware of the philosophical and psychological issues I discuss in Part Two.

You may also be an early career academic who has achieved a teaching position in a university but is not yet sure of the best way to develop your research career in a highly competitive system. If so, I hope that this book has given you some genuine practical help in how to go about the business. It is not for everyone, but an academic research career with all the creativity and freedom of expression that it involves can be a truly fulfilling way to spend your life.

Notes

Introduction

1 In the UK, university life was changed out of all recognition by the introduction of selective research funding by the Research Assessment Exercise (RAE) from 1986 onwards. It has recently been renamed as the Research Excellence Framework (REF). Essentially, it works as follows: at periodic intervals (about every six years), all departments that wish to obtain funding for research work must submit evidence to the relevant funding council to demonstrate both the quantity and especially the quality of research work that has been conducted in the period since the previous assessment in order to secure their level of funding for the next period. Funding depends upon the number of staff engaged in research and the judged quality of the work they do. To those who have much, more will be given. The political motivation behind this was to allow a large expansion in the proportion of the population attending university and reading degrees, while keeping the costs manageable. The previous system of non-selective research funding was considered impractical.

When the RAE was introduced, I believe that government assumed that top researchers would naturally migrate to a relatively small number of elite university departments, leaving the majority of departments to support their degree teaching with scholarship and low-cost research activities. What actually happened was that the quantity and quality of research in British universities improved out of all recognition. Academics are smart people. Given a set of clear criteria in order to obtain both funding and status for their universities, academic managers put in place strong strategies to meet them. Unproductive staff came under great pressure to publish and the very survival of some departments depended upon the RAE ratings they could achieve. Appointing and retaining staff with strong research records became a major priority, leading, for example, to a huge increase in the number of professorial posts in UK psychology departments. Facilities, support and encouragement for research work were all increased. Departments that achieved high ratings reinvested the resulting income in new staff. They also started to appoint staff of related interests in order to build strong research groups, rather than employing staff with a wide range of interests to reflect teaching requirements as had been typical in the days before RAE.

There is no tenure-track system in the UK, so lectureships are normally permanent, barring redundancy. This system makes the appointment of lecturing staff a hazardous business for universities and departments that aspire to high RAE/REF ratings. If you get it wrong, an unproductive member of staff could be with you for a very long time. Hence, the days of appointing people at the point of completion of their PhD's, on the basis of judged potential, are well behind us. I find that I need to explain this to our own research students on a regular basis. No strong department – of the kind you want to work in – will take you on in the UK until you have a track record of high quality publication. So you generally need to spend several years working on short-term postdoctoral research contracts to build up such a record.

1   Scholarship and the origin of ideas

1 This is a quite controversial subjectivist definition of knowledge that you might or might not want to accept. I am not trying to argue that knowledge or understanding is distinctively biological in the manner of the philosopher John Searle (1992). I concede that, in principle, an artificially intelligent system could display knowledge and understanding, although I have yet to see any convincing evidence that such a system has been developed.

2 Presenting the themes of dual process theories in a single table (see also Stanovich, 1999; 2011) contributed to a problem in which some critics attacked a generic 'received' version of dual process theory as though this could apply to all individual accounts. Eventually, we had to show that such criticisms fail, as there is no definitive set of attributes underlying all the theories (Evans & Stanovich, 2013b).

4   Research funding

1 The funding of indirect costs of research grants is something that will vary greatly between countries. In the UK, government research councils now pay the majority (but not all) of indirect costs, but charitable foundations pay only direct costs. However, research grant income is a factor in determining the infrastructure funding through the REF/RAE system, so the overall financial impact on individual universities is complex. There is also a debate as to whether teaching income, raised by student fees, may sometimes be used to subsidise research costs.

6   Collaboration and supervision

1 Unfortunately, we do not live in an ideal world and there are senior academics who may appropriate and use the ideas of young researchers in an unscrupulous manner. If in spite of my cautionary comments you feel that this has happened to you, then it might be best to seek advice of a senior and disinterested colleague in the department before making a formal complaint. It is likely that this will not be the first such occurrence and that his/her colleagues will be aware of it. In this case, it may also be that the individual concerned has a sufficiently powerful position that nothing has been or will be done about it. You may be able to get yourself quietly re-assigned to a different supervisor, but if the practice is endemic to the group you work in, you may need to move on.

2 In UK universities, PhD completion rates were scandalously low 30 years ago. The funding research councils started to insist that students complete within a year of the completion of their funding – that is, typically within four years, given a three year funding period. Although this greatly improved completion times and rates, it seems now to have resulted in a mindset that four years rather than three is the permitted period. Hence, many students end up with an unfunded and stressful year in which they are not able to apply for postdoctoral funding. In the USA, the funding period is normally four years, but with more ancillary duties required. Analogously, five years is often considered a reasonable completion time.

7   Communication of research

1 The successful thriller writer Lee Child has observed that the secret to a good novel is to ask a question at the beginning and answer it at the end. The same applies to journal articles.

2 It is a common observation of experienced researchers that original and ground-breaking research can sometimes be harder to publish. I think this reflects the mindset of busy referees who develop schemas for evaluating papers for particular kinds of journals.

Research that uses familiar methods to investigate familiar issues is hence easy to evaluate. Novel papers, especially from unknown authors, can also cause referees some uncertainty and anxiety. If the journal is prestigious, you do not want to look a fool by recommending publication only to find that the other referees and editor saw all the flaws that you missed! Better to play safe. This can be very frustrating for the researchers if the paper is rejected. In such cases, you just need to persist either with the same editor (if he/she leaves the door open) or with other journals.

9   Statistical inference

1 In some cases, relevant statistics are available and known to decision makers: for example, the divorce rate for those contemplating marriage and the early death rate for cigarette smokers. Although these may influence decision making, they are most unlikely to provide actual probabilities for a number of reasons. The major problem is that the most relevant base rate is never available to an individual. For example, one of my sons complained to me a few years ago about the high car insurance rates for males under 21 years of age. I pointed out the actuarial statistics show that this group have a very high accident rate. His response to this was, 'Yes, but I am a careful driver; why should I be penalised for all those idiot guys I see tearing around the roads?' This was not simply self-delusion, although that is often also involved. For example, if the national divorce rate is 40% it does not follow that your marriage has a 40% chance of failing. It may be higher or lower depending upon the people you are and the exact circumstances. In fact, the circumstances of almost all real world prospects are unique and base rates can never provide more than one source of input to your estimation, which has to be subjective. Your judgement may, of course, also be hopelessly biased and inaccurate.

2 Another way of looking at this, suggested by a reviewer of this manuscript, is that the experimental question in such cases is uninteresting. If you ask an interesting question, according to this perspective, then you cannot have too much power. However, investigating small effects is not usually interesting.

3 There are cases where small effects are important, although this is not normally the case in theoretical research in psychology. Examples would include a weak predictor of performance in an entry test to a large organization, such as the army, where significant costs of training might be saved, or a weak effect of a cheap prophylactic drug that might save many lives.

4 Technically, the power of sample is independent of population size when the sampling is done with replacement, which it never is. In practice, we always sample n *different* members of the population. But the error caused by this is negligible unless the population size is very small. In the studies cited, populations were large compared with samples, so this can be disregarded.

References

Baddeley, A. D. & Hitch, G. J. (1974). Working memory. In G. A. Bower (Ed.), *The psychology of learning and motivation, Volume 8* (pp. 47–90). New York: Academic Press.

Bar-Hillel, M. (1979). The role of sample size in sample evaluation. *Organizational Behavior and Human Performance, 24*, 245–257.

Barbey, A. K. & Sloman, S. A. (2007). Base-rate respect: From ecological validity to dual processes. *Behavioral and Brain Sciences, 30*, 241–297.

Berry, D. C. & Dienes, Z. (1993). *Implicit learning.* Hove, UK: Erlbaum.

Bruner, J. S., Goodnow, J. J., & Austin, G. A. (1956). *A Study of Thinking.* New York: Wiley.

Bueler, R., Griffin, D., & Ross, M. (2002). Inside the planning fallacy: The causes and consequences of optimistic time predictions. In T. Gilovich, D. Griffin, & D. Kahneman (Eds.), *Heuristics and biases: The psychology of intuitive judgement* (pp. 250–270). Cambridge: Cambridge University Press.

Casscells, W., Schoenburger, A., & Graboys, T. B. (1978). Interpretation by physicians of clinical laboratory results. *New England Journal of Medicine, 299*, 999–1001.

Chapman, L. J. (1967). Illusory correlation in observational report. *Journal of Verbal Learning and Verbal Behavior, 73*, 193–204.

Christensen, L. (2006). *Experimental methodology (10th Edition).* Needham Heights, MA: Allyn & Bacon.

Clark, H. H. (1969). Linguistic processes in deductive reasoning. *Psychological Review, 76*, 387–404.

Cohen, J. (1962). The statistical power of abnormal-social psychological research. *Journal of Abnormal and Social Psychology, 65*, 145–153.

Cosmides, L. & Tooby, J. (1994). Beyond intuition and instinct blindness: Toward an evolutionary rigorous cognitive science. *Cognition, 50*, 41–77.

Cosmides, L. & Tooby, J. (1996). Are humans good intuitive statisticians after all? Rethinking some conclusions from the literature on judgment under uncertainty. *Cognition, 58*, 1–73.

Cummins, D. D. (2004). The evolution of reasoning. In R. J. Sternberg (Ed.), *The nature of reasoning* (pp. 339–374). Cambridge: Cambridge University Press.

Dennett, D. (1991). *Consciousness explained.* London: Allen Lane.

Doherty, M. E., Chadwick, R., Garavan, H., Barr, D., & Mynatt, C. R. (1996). On people's understanding of the diagnostic implications of probabilistic data. *Memory & Cognition, 24*, 644–654.

Doherty, M. E. & Kurz, E. M. (1996). Social judgement theory. *Thinking & Reasoning, 2*, 109–140.

Dunbar, K. N. (2002). Understanding the role of cognition science. In P. Carruthers, S. Stich, & M. Siegal (Eds.), *The cognitive basis of science* (pp. 154–170). Cambridge: Cambridge University Press.

Evans, J. St. B. T. (1977). Toward a statistical theory of reasoning. *Quarterly Journal of Experimental Psychology, 29*, 297–306.

Evans, J. St. B. T. (1982). *The psychology of deductive reasoning*. London: Routledge.

Evans, J. St. B. T. (1989). *Bias in human reasoning: Causes and consequences*. Brighton: Erlbaum.

Evans, J. St. B. T. (1992). Bias in thinking and judgement. In M. T. Keane & K. J. Gilhooly (Eds.), *Advances in the psychology of thinking (Volume 1)* (pp. 95–125). Harvester-Wheatsheaf: UK.

Evans, J. St. B. T. (1995). Implicit learning, consciousness and the psychology of thinking. *Thinking & Reasoning, 1*, 105–118.

Evans, J. St. B. T. (2002). Logic and human reasoning: An assessment of the deduction paradigm. *Psychological Bulletin, 128*, 978–996.

Evans, J. St. B. T. (2003). In two minds: Dual process accounts of reasoning. *Trends in Cognitive Sciences, 7*, 454–459.

Evans, J. St. B. T. (2004). History of the dual process theory of reasoning. In K. I. Manktelow & M. C. Chung (Eds.), *Psychology of reasoning: Theoretical and historical perspectives* (pp. 241–266). Hove, UK: Psychology Press.

Evans, J. St. B. T. (2007a). *Hypothetical thinking: Dual processes in reasoning and judgement*. Hove, UK: Psychology Press.

Evans, J. St. B. T. (2007b). On the resolution of conflict in dual-process theories of reasoning. *Thinking & Reasoning, 13*, 321–329.

Evans, J. St. B. T. (2008). Dual-processing accounts of reasoning, judgment and social cognition. *Annual Review of Psychology, 59*, 255–278.

Evans, J. St. B. T. (2010). *Thinking twice: Two minds in one brain*. Oxford: Oxford University Press.

Evans, J. St. B. T. (2014). Reasoning, biases and dual processes: The lasting impact of Wason (1960). *Quarterly Journal of Experimental Psychology, 68*, xx.

Evans, J. St. B. T. & Ball, L. J. (2010). Do people reason on the Wason selection task?: A new look at the data of Ball et al. (2003). *Quarterly Journal of Experimental Psychology, 63*, 434–441.

Evans, J. St. B. T., Barston, J. L., & Pollard, P. (1983). On the conflict between logic and belief in syllogistic reasoning. *Memory & Cognition, 11*, 295–306.

Evans, J. St. B. T. & Bradshaw, H. (1986). Estimating sample size requirements in research design: A study of intuitive statistical judgement. *Current Psychological Research and Reviews, 5*, 10–19.

Evans, J. St. B. T., Clibbens, J., Cattani, A., Harris, A., & Dennis, I. (2003). Explicit and implicit processes in multicue judgement. *Memory & Cognition, 31*, 608–618.

Evans, J. St. B. T., Clibbens, J., & Rood, B. (1996). The role of implicit and explicit negation in conditional reasoning bias. *Journal of Memory and Language, 35*, 392–409.

Evans, J. St. B. T. & Curtis-Holmes, J. (2005). Rapid responding increases belief bias: Evidence for the dual-process theory of reasoning. *Thinking & Reasoning, 11*, 382–389.

Evans, J. St. B. T. & Dusoir, A. E. (1977). Proportionality and sample size as factors in intuitive statistical judgement. *Acta Psychologica, 41*, 129–137.

Evans, J. St. B. T., Handley, S. J., & Over, D. E. (2003). Conditionals and conditional probability. *Journal of Experimental Psychology: Learning Memory and Cognition, 29*, 321–335.

Evans, J. St. B. T., Handley, S. J., Perham, N., Over, D. E., & Thompson, V. A. (2000). Frequency versus probability formats in statistical word problems. *Cognition, 77*, 197–213.

Evans, J. St. B. T. & Johnson-Laird, P. N. (2003). Editorial Obituary: Peter Wason 1924–2003. *Thinking & Reasoning, 9*, 177–184.

Evans, J. St. B. T. & Lynch, J. S. (1973). Matching bias in the selection task. *British Journal of Psychology, 64*, 391–397.

Evans, J. St. B. T., Newstead, S. E., & Byrne, R. M. J. (1993). *Human reasoning: The psychology of deduction*. Hove, UK: Erlbaum.

Evans, J. St. B. T. & Over, D. E. (1996). *Rationality and reasoning*. Hove: Psychology Press.

Evans, J. St. B. T. & Over, D. E. (2004). *If*. Oxford: Oxford University Press.

Evans, J. St. B. T. & Over, D. E. (2013). Reasoning to and from belief: Deduction and induction are still distinct. *Thinking and Reasoning, 19*, 267–283.

Evans, J. St. B. T., Over, D. E., & Handley, S. J. (2005). Supposition, extensionality and conditionals: A critique of Johnson-Laird & Byrne (2002). *Psychological Review, 112*, 1040–1052.

Evans, J. St. B. T. & Stanovich, K. E. (2013a). Theory and metatheory in the study of dual processing: A reply to comments. *Perspectives on Psychological Science, 8*, 263–271.

Evans, J. St. B. T. & Stanovich, K. E. (2013b). Dual process theories of higher cognition: Advancing the debate. *Perspectives on Psychological Science, 8*, 223–241.

Evans, J. St. B. T. & Wason, P. C. (1976). Rationalisation in a reasoning task. *British Journal of Psychology, 63*, 205–212.

Feynman, R. P. (1985). *Surely you are joking Mr Feynman!* Reading: Vintage.

Fiddick, L., Cosmides, L., & Tooby, J. (2000). No interpretation without representation: the role of domain-specific representations and inferences in the Wason selection task. *Cognition, 77*, 1–79.

Fischhoff, B. (1982). For those condemned to study the past: Heuristics and biases in hindsight. In D. Kahneman, P. Slovic, & A. Tversky (Eds.), *Judgement under uncertainty: Heuristics and biases* (pp. 335–351). Cambridge: Cambridge University Press.

Fodor, J. (2001). *The mind doesn't work that way*. Cambridge, MA: MIT Press.

Fugelsang, J. A., Stein, C. B., Green, A. E., & Dunbar, K. N. (2004). Theory and data interactions of the scientific mind: Evidence from the molecular and cognitive laboratory. *Canadian Journal of Experimental Psychology, 58*, 86–95.

Gigerenzer, G. (2002). *Reckoning with risk*. London: Penguin Books.

Gigerenzer, G. & Murray, D. J. (1987). *Cognition as intuitive statistics*. Hillsdale, NJ: Erlbaum.

Gilovich, T., Griffin, D., & Kahneman, D. (2002). *Heuristics and biases: The psychology of intuitive judgement*. Cambridge: Cambridge University Press.

Gorman, M. E. (1995). Confirmation, disconfirmation and invention: The case of Alexander Graham Bell and the telephone. *Thinking & Reasoning, 1*, 31–54.

Gregory, R. L. (1963). Distortion of space as inappropriate constancy. *Nature, 199*, 678–680.

Gregory, R. L. (1990). *Eye and brain (4th edition)*. Princeton, NJ: Princeton University Press.

Griffin, D. & Tversky, A. (1992). The weighting of evidence and the determinants of confidence. *Cognitive Psychology, 24*, 411–435.

Harries, C., Evans, J. St. B. T., & Dennis, I. (2000). Measuring doctors' self-insight into their treatment decisions. *Journal of Applied Cognitive Psychology, 14*, 455–477.

Howson, C. & Urbach, P. (2006). *Scientific reasoning: The Bayesian approach (3rd edition)*. Chicago: Open Court.

Janis, I. (1982). *Victims of groupthink (2nd edition)*. Boston: Houghton Mifflin.

Johnson-Laird, P.N. (1983). *Mental models*. Cambridge: Cambridge University Press.

Johnson-Laird, P.N. & Bara, B.G. (1984). Syllogistic inference. *Cognition, 16*, 1–61.

Johnson-Laird, P.N. & Byrne, R.M.J. (1991). *Deduction*. Hove & London: Erlbaum.

Johnson-Laird, P.N. & Byrne, R.M.J. (2002). Conditionals: A theory of meaning, pragmatics and inference. *Psychological Review, 109*, 646–678.

Johnson-Laird, P.N. & Savary, F. (1999). Illusory inferences: A novel class of erroneous deductions. *Cognition, 71*, 191–299.

Johnson-Laird, P.N. & Wason, P.C. (1970). A theoretical analysis of insight into a reasoning task. *Cognitive Psychology, 1*, 134–148.

Kahneman, D., Slovic, P., & Tversky, A. (1982). *Judgment under uncertainty: Heuristics and biases*. Cambridge: Cambridge University Press.

Kahneman, D. & Tversky, A. (1972). Subjective probability: A judgment of representativeness. *Cognitive Psychology, 3*, 430–454.

Kahneman, D. & Tversky, A. (1982). On the study of statistical intuition. *Cognition, 12*, 325–326.

Klauer, K.C., Musch, J., & Naumer, B. (2000). On belief bias in syllogistic reasoning. *Psychological Review, 107*, 852–884.

Klauer, K.C., Stahl, C., & Erdfelder, E. (2007). The abstract selection task: New data and an almost comprehensive model. *Journal of Experimental Psychology: Learning, Memory, and Cognition, 33*, 680–703.

Klayman, J. (1995). Varieties of confirmation bias. *The Psychology of Learning and Motivation, 32*, 385–417.

Klayman, J. & Ha, Y.W. (1987). Confirmation, disconfirmation and information in hypothesis testing. *Psychological Review, 94*, 211–228.

Koehler, J.J. (1996). The base rate fallacy reconsidered: Descriptive, normative and methodological challenges. *Behavioral and Brain Sciences, 19*, 1–53.

Koonin, E.K. & Wolf, Y. (2009). Is evolution Darwinian or/and Lamarckian? *Biology Direct, 4:42*, doi 10.1186/1745–6150–4–42.

Kuhn, T.S. (1970). *The structure of scientific revolutions (2nd edition)*. Chicago: University of Chicago Press.

Lewis, D. (1981). Causal decision theory. *Australasian Journal of Philosophy, 59*, 5–30.

Lord, C., Ross, L., & Lepper, M.R. (1979). Biased assimilation and attitude polarisation: The effect of prior theories on subsequently considered evidence. *Journal of Personality and Social Psychology, 37*, 2098–2109.

Manktelow, K.I. (2012). *Thinking and reasoning*. Hove, UK: Psychology Press.

McClelland, J.M. & Rumelhart, D.E. (1985). Distributed memory and the representation of general and specific information. *Journal of Experimental Psychology, 114*, 159–188.

McClelland, J.M. & Rumelhart, D. (1986). *Parallel distributed processing: Experiments in the microstructure of cognition (Volume 2): Psychological and biological models*. Cambridge, MA: MIT Press.

Miller, G.A., Gallanter, E., & Pribram, K. (1960). *Plans and the structure of behavior*. New York: Holt.

Murzi, M. (2001). Logical positivism. *The internet encyclopedia of philosophy*, http://www.iep.utm.edu/logpos.htm

Neisser, U. (1967). *Cognitive psychology*. New York: Appleton.

Neisser, U. (1976). *Cognition and reality*. San Francisco: Freeman.

Newell, A. (1973). You can't play 20 questions with nature and win. In W.G. Chase (Ed.), *Visual information processing* (pp. 283–308). New York: Academic Press.

Newell, A. & Simon, H.A. (1972). *Human problem solving*. Englewood Cliffs, NJ: Prentice-Hall.

Nisbett, R. & Ross, L. (1980). *Human inference: Strategies and shortcomings of social judgement*. Englewood Cliffs, NJ: Prentice-Hall.

Nisbett, R.E. & Wilson, T.D. (1977). Telling more than we can know: Verbal reports on mental processes. *Psychological Review, 84*, 231–295.

Oaksford, M. & Chater, N. (2007). *Bayesian rationality: The probabilistic approach to human reasoning*. Oxford: Oxford University Press.

Over, D.E. (2003). From massive modularity to metarepresentation: The evolution of higher cognition. In D.E. Over (Ed.), *Evolution and the psychology of thinking: The debate* (pp. 121–144). Hove, UK: Psychology Press.

Pearl, J. (2000). *Causality: Models, reasoning and inference*. Cambridge: Cambridge University Press.

Pollard, P. & Richardson, J.T.E. (1987). On the probability of making type 1 errors. *Psychological Bulletin, 102*, 159–163.

Popper, K.R. (1959). *The logic of scientific discovery*. London: Hutchinson.

Popper, K.R. (1962). *Conjectures and refutations*. London: Hutchinson.

Potts, G.R. & Scholz, K.W. (1975). The internal representation of three-term series problems. *Journal of Verbal Learning and Verbal Behavior, 11*, 727–740.

Reber, A.S. (1993). *Implicit learning and tacit knowledge*. Oxford: Oxford University Press.

Reber, R. (2004). Availability. In R. Pohl (Ed.), *Cognitive illusions* (pp. 147–164). Hove, UK: Psychology Press.

Rosenthal, R. (1993). *Meta-analytic procedures for social research*. London: Sage.

Schneider, W. & Shiffrin, R.M. (1977). Controlled and automatic human information processing I: Detection, search and attention. *Psychological Review, 84*, 1–66.

Schroyens, W. & Schaeken, W. (2004). Guilt by association: On iffy propositions and the proper treatment of mental models theory. *Current Psychology Letters* (http://cpl.revues.ord/document411.html), 12.

Searle, J. (1992). *The rediscovery of mind*. Cambridge, MA: MIT Press.

Sedlmeier, P. & Gigerenzer, G. (1989). Do studies of statistical power have an effect on the power of studies? *Psychological Bulletin, 105*, 309–316.

Shiffrin, R.M. & Schneider, W. (1977). Controlled and automatic human information processing II: Perceptual learning, automatic attending and a general theory. *Psychological Review, 84*, 127–189.

Simon, H.A. (1982). *Models of bounded rationality*. Cambridge, MA: MIT Press.

Sloman, S.A. (1996). The empirical case for two systems of reasoning. *Psychological Bulletin, 119*, 3–22.

Sloman, S.A. (2005). *Causal models*. Oxford: Oxford University Press.

Stanovich, K.E. (1999). *Who is rational? Studies of individual differences in reasoning*. Mahway, NJ: Lawrence Erlbaum Associates.

Stanovich, K.E. (2004). *The robot's rebellion: Finding meaning in the age of Darwin*. Chicago: University of Chicago Press.

Stanovich, K.E. (2011). *Rationality and the reflective mind*. New York: Oxford University Press.

Stanovich, K.E. & West, R.F. (2003). Evolutionary versus instrumental goals: How evolutionary psychology misconceives human rationality. In D.E. Over (Ed.), *Evolution and the psychology of thinking* (pp. 171–230). Hove, UK: Psychology Press.

Sternberg, R. J. (1980). Representation and process in linear syllogistic reasoning. *Journal of Experimental Psychology: General, 109,* 119–159.

Tanner, W. P. & Swets, J. A. (1954). A decision-making theory of visual detection. *Psychological Review, 11,* 167–187.

Thompson, V. A., Evans, J. St. B. T., & Campbell, J. I. D. (2013). Matching bias on the selection task: It's fast and feels good. *Thinking and Reasoning, 19,* 431–452.

Thompson, W. C. & Schumann, E. J. (1987). Interpretation of statistical evidence in criminal trials: The prosecutor's fallacy and the defense attorney's fallacy. *Law and Human Behavior, 11,* 167–187.

Tversky, A. & Kahneman, D. (1971). The belief in the 'law of small numbers'. *Psychological Bulletin, 76,* 105–110.

Tversky, A. & Kahneman, D. (1973). Availability: A heuristic for judging frequency and probability. *Cognitive Psychology, 5,* 207–232.

Wagenaar, W. A. (1988). *Paradoxes of gambling behaviour.* Hove & London: Erlbaum.

Wason, P. C. (1960). On the failure to eliminate hypotheses in a conceptual task. *Quarterly Journal of Experimental Psychology,* 12–40.

Wason, P. C. (1966). Reasoning. In B. M. Foss (Ed.), *New horizons in psychology 1* (pp. 106–137). Harmondsworth: Penguin.

Wason, P. C. (1968). On the failure to eliminate hypotheses: A second look. In P. C. Wason & P. N. Johnson-Laird (Eds.), *Thinking and reasoning* (pp. 165–174). Harmondsworth: Penguin.

Wason, P. C. (1972). In real life negatives are false. *Logique et Analyse, 19,* 19–38.

Wason, P. C. (1995). Creativity in research. In S. E. Newstead & J. St. B. T. Evans (Eds.), *Perspectives on thinking and reasoning: Essays in honour of Peter Wason* (pp. 287–301). Hove, UK: Lawrence Erlbaum Associates.

Wason, P. C. & Brooks, P. G. (1979). THOG: The anatomy of a problem. *Psychological Research, 41,* 79–90.

Wertheimer, M. (1961). *Productive thinking.* London: Tavistock.

Subject Index

Author Index